AC Power Interference Handbook

2nd Edition, Revised

by

Marv Loftness

KB7KK

AC Power Interference Handbook

New Insights into the Causes, Effects, Locating, and Correction of Power-Line and Other Electrical Interference.

by ***Marv Loftness***

ISBN: 0-9653760-3-6

Published by Percival Technology
P.O. Box 4172, Tumwater, WA 98501

Distributed by ARRL
225 Main St.
Newington, CT 06111-1494
1-(888) 277-5289

Printed by Bookmasters, Inc.
2541 Ashland Rd, P.O. Box 2139
Mansfield, OH 44905
1-(419)-589-5100

Cover Design by
Washington Media Services
P.O. Box 7184
Olympia, WA 98507
1-(360)-754-4543

TABLE OF CONTENTS

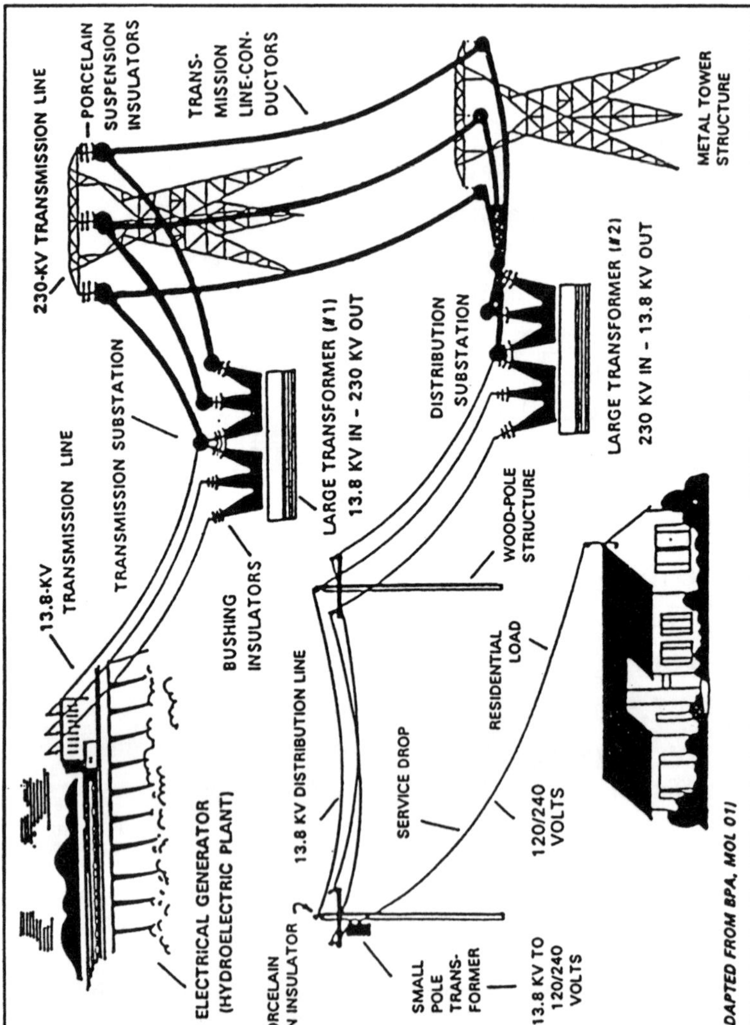

Overview of an AC Power Sytem

AN OVERVIEW OF THE AC POWER SYSTEM

The AC Power System. If it were not for AC power systems, there would be no need for us to be concerned about AC power interference. We do have such systems, of course, and *many* of them. An example system is shown in the figure and this will prepare you for some of our power-line interference discussions.

The Generator, the source of our power, operates at a relatively low voltage, here 13.8 thousand volts (13.8 kV). Huge transformers at the generator site boost this voltage to transmission line voltages, such as 230 kV.

The Transmission Lines are the higher-voltage power lines. Their high voltage allows the electric power to be carried efficiently over long distances, from the generating facilities such as hydro-electric dams and coal or gas-fired plants, to substations near power consumers. Transmission lines using 69 kV, 115 kV, 230 kV, and 500 kV are common, although lines with many other voltage ratings are in use.

The Distribution Lines of the local utilities use lower voltage lines to bring power from substations to homes or commercial buildings. These operate at 12 kV through 35 kV typically, although many other voltages are also used. These are sometimes called *feeder* lines.

The Pole Transformer near your house drops the distribution voltage down to that used in a home or in other buildings. This is typically 220/120 volts. These pole-to-house lines are called service lines.

Note that the transmission lines shown are on metal towers and the distribution lines are on wood poles. However, transmission lines are often found on high wood poles. Distribution lines are usually on wood poles but sometimes on metal structures. Concrete and composite structures are sometimes used for either distribution or transmission lines.

Chapter One

CAUSES AND EFFECTS

Fig. 1. A. One common Cause.   **B.** One common Effect.

Introduction

Fig. 1 shows a very common *cause* **A**, of power-line noise and a very common *effect* **B**, television interference. The cause here is arcing in the hardware holding two porcelain insulators together in a vertical "insulator string"on a distribution line.

The bright light in the center of photo **A** is caused by a small arc–in a gap of less than 0.1 inch. The author classifies this as a spark. It is a good example of a *sparking-noise source.*

As a power company investigator the author finds that the customers, when experiencing interference of one sort or another, will often call this interference "power-line noise." They are sometimes right, but often wrong.

Here we have referred to the complainant as a *customer*. This is usually the case, whether a customer of the power company or a customer of the cable-TV company.

In A of Fig. 1, the arc is quite small. The differences between a spark and a large power-arc are pronounced, but either may be called an arc. These differences will be illustrated.

Complaints to the Power Company

Most complaints to a power company are TV watchers, radio amateurs, AM radio listeners, cable companies, telephone companies, and communication companies or agencies. Occasionally, computer operators complain. All of these may be affected in various ways, but most complain about power-line noise. However, a good share of the complaints received by a power company are found to be caused by "non-power-line" noise sources.

Often at a customer's residence or location it will be noted that radio-frequency interference (RFI) from *more than one noise source* is being radiated and/or conducted into his location and into his receiving equipment.

If you are a customer, have an interference problem, and can determine that one or more of the noises bothering you *is* a power-line source, you should, of course, call the power company for help. Chapter 4 will help you determine if the "noise" (be it radio or television interference) is coming from the power line or elsewhere.

To define power-line interference, the IEEE (Institute of Electronic and Electrical Engineers) prefers the term EMI, electromagnetic interference, rather than RFI. The author doesn't use this because, for one thing, if I introduce myself as as a power company EMI investigator, the complainant may want to get into a discussion of the biological effects of power lines. This is of interest, but is a distraction from the noise investigation at hand. (Incidentally, the term often used by the general public for biological effects is EMF, by which they mean *electric* and *magnetic* fields or *electromagnetic* fields)

Various Power-Line Noises

Corona noise as well as sparking is found on distribution lines, but it causes very weak radio interference on these lower-voltage lines and may generally be disregarded at distribution-voltage levels. There are exceptions however, and corona will be discussed in Chapter 2.

Telephone noise is another type of noise and is discussed in Chapter 13. Telephone noise resulting from nearby power lines is the result of magnetic induction and is NOT RF radiation or RF induction. This noise is troublesome to the telephone companies, to railroad companies, and to other "earth-bound" communication services, but it generally does not disturb radio, television, or radio amateur operations. In the telephone noise chapter, the term "noise" does not mean RFI as in the other chapters of this handbook.

> ***Audio noise*** (AN) from a power line is also an occasional complaint–this usually is noise from power lines of extra-high-voltage (EHV), 345 kV or higher. Corona on these lines is the main cause. However AN also results from aeolian conductor vibration. When such audio noise is discussed it will be called *AN* to distinguish it from the "*noise*" which is used interchangeably with "*RFI*."

Some effects of sparking noise and their relation to 60 Hz are seen in Fig. 2.

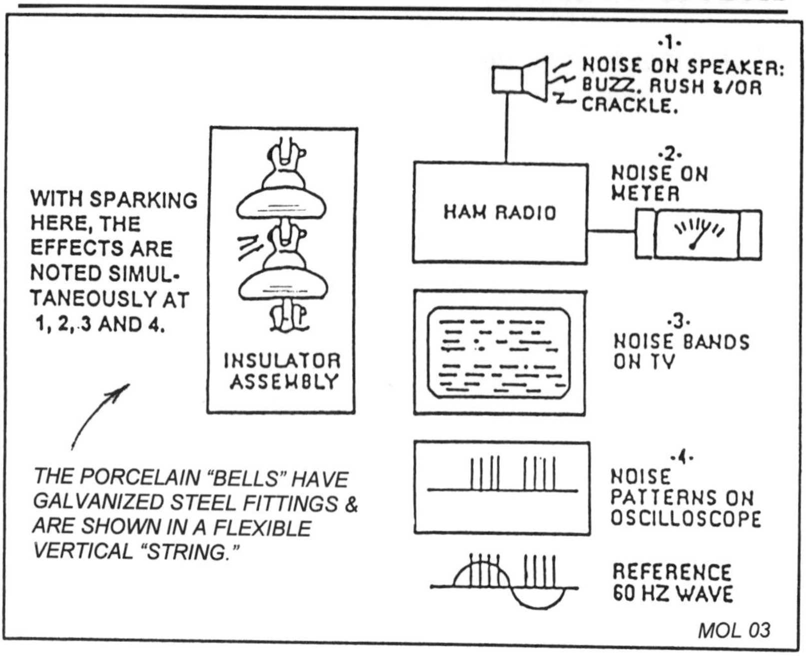

Fig. 2. Various Effects, Power-Line Sparking.

The figure indicates four effects of power-line sparking: noise heard on a radio receiver; and noise seen on a voltage meter, a TV or an oscilloscope. These observations, at 1, 2, 3, and 4 occur simultaneously.

In the above, the sparking is taking place between the tongue and clevis of the insulator units, and if we had good electrical connection between these parts we would have no sparking. This will be expanded upon in later discussions.

[If you hear a buzz or a hum associated with a single sparking source, what your are hearing is the pulse-group repetition rate, and this is occurring at 120 Hz, if the pattern is like that seen in 4 of Fig. 2. If full rectification is taking place (as in Fig. 9-A on page 1-13) you are hearing a frequency of 60-Hz. However, you may not recognize by ear the frequency difference between a 60-Hz buzz and a 120-Hz buzz, without a direct comparison.

Sparking also occurs frequently on transmission lines, particularly on those of wood-pole construction. An example is shown in Fig. 3. Here we have a 115-kV line with porcelain post insulators, and sparking noise is being generated in the hardware clamping the line-conductor to the post insulator.

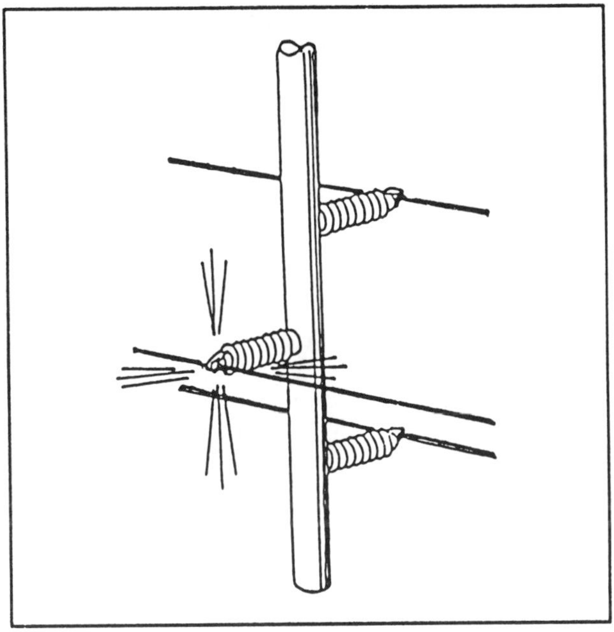

Fig. 3. RFI Radiation Source on Transmission Line. Distribution lines also often use post insulators, sometimes with trunnion clamps, but usually these are the tie-wire type.

In the above illustration, the radio-frequency *interference caused by sparking* is radiating in all directions, into residences near the power line, into the surrounding hills–and into the space above. In addition to this radiation, the noise is propagated along the line in both directions and reradiating from many power-line spans and poles. This reradiation may occur miles from the source on a *transmission* line. On *distribution* lines, such RF line-conduction and *reradiation* is usually limited to a mile or less, sometimes to only a few poles.

The sparking of Fig. 3 is occurring between the trunnion pins and the conductor-clamp because of a loose end- pivot-bolt.

Fig 4. shows details of the mounting. Here a line-conductor in its clamp is about to be mounted in its post-insulator end-trunnion.

You may find that this problem only occasionally occurs. However, you also may find this pivot-pin source occurring on many structures in a section of line as a result of incorrect line construction or maintenance.

Fig. 4.

Pre-Assembly View: Conductor Clamp to Post Insulator Trunnion.

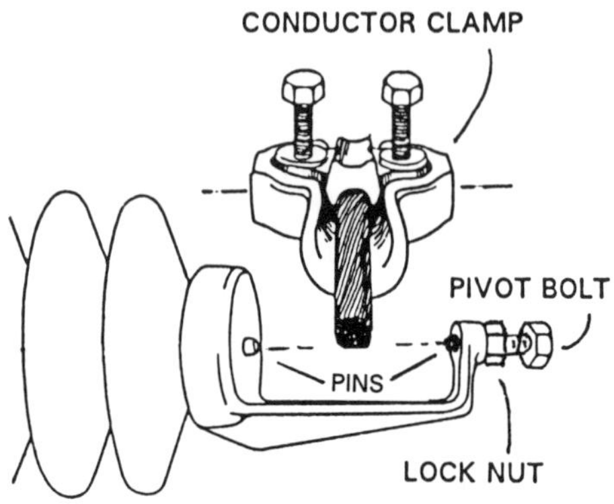

In the above the author uses a somewhat unconventional term: "reradiation." Some experts think we should refer to such as "standing waves," or reflections, but this would confuse our discussions. From power-line sparking we have very broad-band reradiation, which is loaded with many frequencies, whereas a standing wave is thought of as being a single frequency.

We have radiation at a source location, and, as a result, *reradiation* at other locations on the power-line conductor. You will appreciate this distinction when you are locating sources, or when you are describing the locating procedure to others We say, then, that we have *radiation, conduction*, and *reradiation* of noise from or on a power-line.

Sparking Circuit on a Power Line

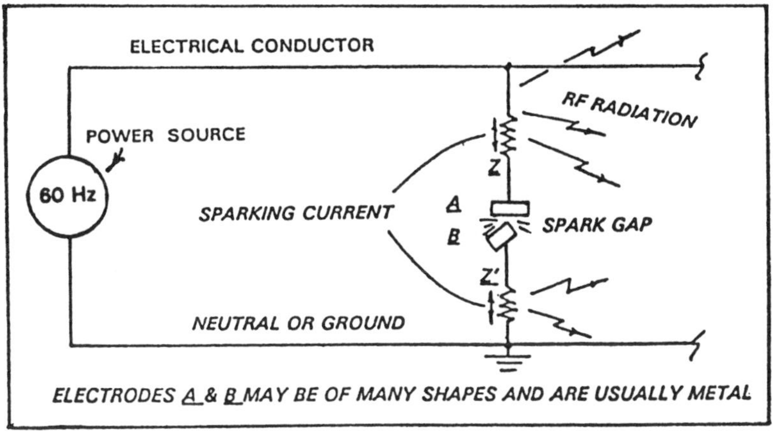

Fig. 5. The Sparking Circuit on a Power Line--Simplified.

When the air breaks down in the gap of a loose power-line connection as a result of voltage stress, current will flow through this gap in a sudden discharge–it is a lighting stroke in miniature! This current causes an immediate voltage drop across the resistance (the series impedance Z+Z') of the sparking circuit. The current pulses resulting are shown in the oscilloscope photo of Fig. 6A and the resulting RF-noise pulses are shown Fig. 6B.

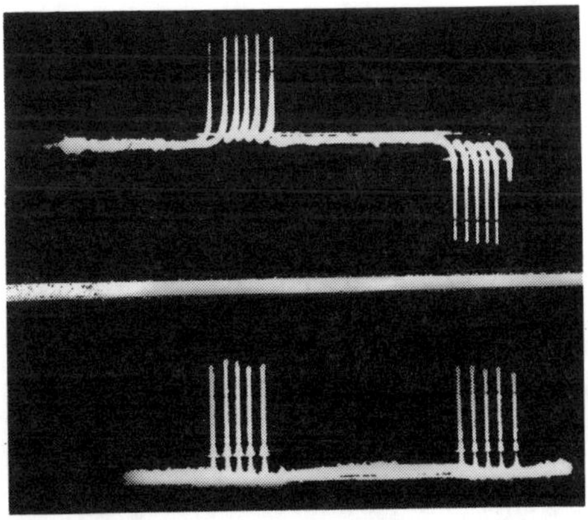

Fig 6A. Gap Current Pulses*

Fig 6B. RF Noise-Voltage Pulses after Detection.

*The voltage drops across resistor z' (Fig. 5), seen as spikes on the oscilloscope, indicate current pulses, as in Fig. 6A.

With a lowered voltage at the gap, resulting from the current flow through the impedances, the spark "goes out." However, the voltage immediately builds up from the charging source (the power-line voltage), and another spark fires.

This rapid firing and rapid dying-out repeats itself for as long as there is sufficient voltage available during each half-cycle of the 60-Hz voltage wave (again referring to Fig. 5).

The sharp current pulses (spikes) shown in Fig. 6 generate *sharp radio-frequency noise voltage* spikes which cause radiation from the conductor or other metal parts associated with the sparking gap.

The detector action in the RF receiver or field strength meter used causes *any noise* pulses received to show up on the scope as *positive* pulses on *both half-cycles* of the 60 Hz. This is shown in Fig. 6B and in Fig. 7.

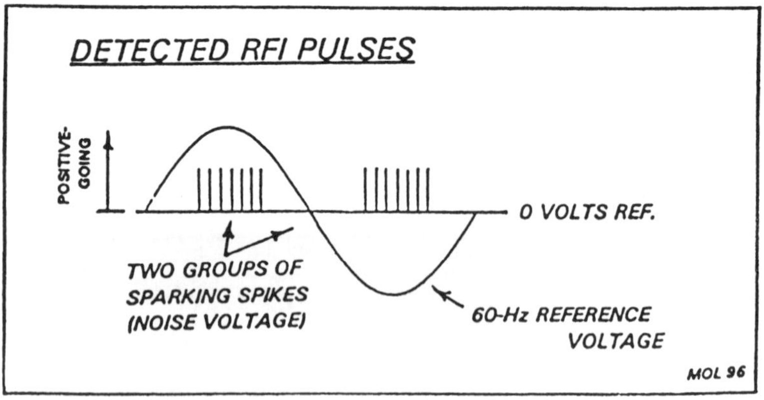

Fig. 7. Detected RFI Noise Spikes Caused by Sparking-Current Spikes. [NRECA Ref. 1].

Sparking Consequences

The rise time of the disruptive sparking current pulse in a gap is a phenomenal 2 nanoseconds, that is 2 billionths of a second [Janischewskyj Ref. 2]. This extremely sharp current discharge, breaking down the air in a gap, *really shakes up the electromagnetic environment!* This causes RF (radio frequency) noise radiation over an extremely wide frequency range from below 1 MHz to above 1000 MHz.

The broad-band noise generated is radiated from *any metal* associated with the sparking "gap source." The metal provides an antenna effect which is essential to the effective "broadcasting" of the RF noise, and the metal objects immediately associated with the gap we call the "gap electrodes."

As an example, one of these electrodes may be a power-line conductor, and the other gap electrode may be a loose connection clamp (a hot-line clamp) on the conductor.

We usually think of the electrodes as being metal, but one of them might be a charged insulating body such as porcelain, polymer or other material.

> By *"radio frequencies"* we mean more than just those of "broadcast radio," and we include TV, shortwave, and amateur radio frequencies; also included are other frequencies which carry information via what we used to call the "airwaves" through empty space.

Referring to the above box, it has been said that in the 1930's many "experts" believed that radio waves traveled via "ether" in the air. No one could define the ether, and the theory was finally given up. Further, these "air waves" do not even require air. The air actually impedes radio waves to some extent and radio frequencies travel more easily in empty space (as do the other frequencies in the electromagnetic spectrum). Our atmosphere is important to global long-distance radio communications, however, as it reflects waves to permit them to "bend" around earth's circumference to a considerable extent, depending on wavelength and other factors. Much of our long-distance radio now is via satellite, however, which does not require any atmosphere.

Getting back to our main subject which is power-line sparking noise:

An interesting fact is that sparking/arcing also causes audio and super audio (ultrasonic) noise. This is observed only close to the sparking source. In the locating of sources, we *can* take advantage of this by the *ultrasonic pinpointing sources* with the proper detector.

In the upper portion of the electromagnetic spectrum, we get into the infra red and other light frequencies. These also are generated by sparks and arcs but are not useful frequencies for our rapid locating system, and they are not a subject of this handbook.

For ultrasonic (super-sound) detection we are concerned with *non-radio* frequencies, such as 20,000 to 40,000 cycles-per-second (cps). The Radar Engineers (RE) Model 247 hotstick sniffer has an ultrasonic probe that detects noise at 40,000 cps, which is above the range of human hearing (where the upper limit is about 16,000 cps). This probe is often used for final pinpointing of a sparking source. [Note that the author here is not using *Hertz (Hz)* and rather uses *cps* or "*cycles*" to distinguish ultrasonic generation from broad-band *RF* generation.]

In general, the *audio sound* we can hear is quite weak even at the sparking source, but power company linemen working near a source will sometimes hear it.

The Spark vs the Arc

A daylight view of a fused-cutout switch is shown in Fig. 8 (over). This cutout is then shown being opened in the dark, with a hot-stick tool. Here you see the stretching of the arc, to break it, during the opening of the switch. If the serviceman stopped the opening of the switch at the point shown, the flaming would continue and the switch would burn up.

> ***The differences between the power arc and the spark are*** pronounced. For one thing, the power arc, as shown in Fig. 8-B, can be stretched to many times its starting length, whereas a spark cannot; also, the arc has a continually conductive *ionized channel* through and between the positive and negative half-cycles of the 60-Hz voltage wave, whereas the spark does not. This ionized channel is sometimes called a *plasma* channel.

In this book, any arc (small or large size) that *produces noise patterns* on the scope, similar to those in this chapter, is considered *sparking*, and to the author, these patterns *define* sparking and RFI.

"Microsparking" and "gap-noise," however, are terms that are sometimes used to mean sparking. "Microsparking" is not used by the author because the gaps involved are sometimes not so *micro*–they may be well over 1/10th inch across. Also "gap-noise" is not used because larger *power arcs* also take place in gaps. They may cause serious damage, but are short-lived, and therefore do not concern us from an RFI-locating standpoint.

Sparking spikes most frequently occur in groups, rather than singly, with one group *per each half-cycle.* Although individual *groups* are short-lived, that is, less than one-half cycle each, these groups may continually recur, doing so for each power-line voltage alternation. A particular sparking source may continue for many minutes and sometimes hours (or even days) at a particular site on the power line.

A.

B.

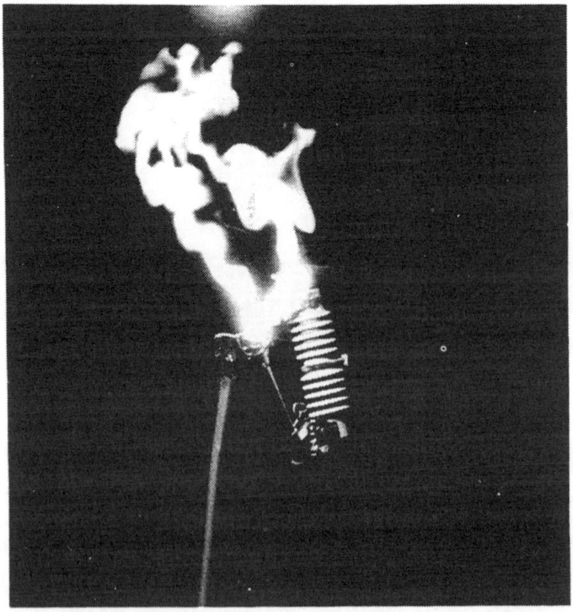

Fig. 8. **A.** Hinged-Fuse Cartridge (F) Fully Engaged.
B. Hinged-Fuse Cartridge Partially Opened.

Sparking Pattern Variations

Rectification and partial rectification in the spark gap contributes *to the great variety of patterns* you will observe on on your scope from a power line. Such rectification on an AC line means that sparking current flows through the gap more readily in one direction than the other–for example, from the positive side of the gap to the negative–and we might see a noise pattern on the on the positive portion of the half-cycle only, as in Fig. 9A.

The rectification shown may be caused by corrosion in the gap, a sharp point on one of the gap electrodes or for other mysterious reasons. We know that it does occur, and frequently. Tests by the author, indicate this rectification in a spark gap takes place on the *negative* half-cycle of the AC 60-Hz voltage wave. In other words, the pulse group on the negative half-cycle is eliminated or reduced.

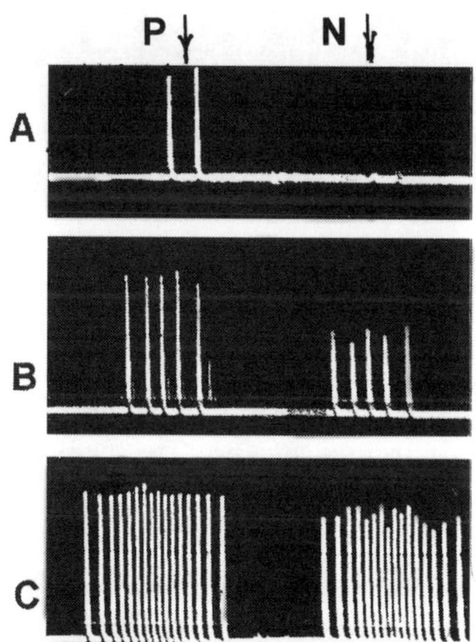

Fig. 9. Different Sparking Patterns You Might Observe on a Power Line. For the three patterns, P is the positive half cycle and N is the negative half cycle.

Full rectification is shown in A and partial rectification is noticeable in B and C. [For all displays the electron beam in the oscilloscope is sweeping across the screen from left to right, at 60-times per second (a 60-Hz rate), and is being deflected vertically by each noise-voltage spike. [Such displays are thanks to Tektronix and other oscilloscope designers].

Multiple sources are frequently seen and, fortunately for the person *locating* sources, the different sources observed have discernibly different patterns, as seen on an oscilloscope. For example, one source pattern may have a different pattern of rectification, be more rectangular or more rounded, or have a different number of spikes per group. One group may have a different voltage levels or be out-of-phase (displaced horizontally on the scope screen with reference to another group patterns as shown in Fig. 10).

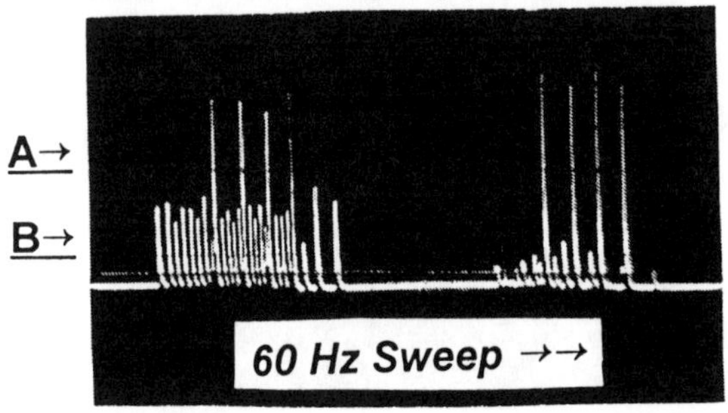

Fig. 10. Example of Scope Display with Two Sources.

It is seen that the ***B*** group patterns are denser and have about half the peak-voltage level of the ***A*** patterns, and that the ***A*** groups occur later in time.

These patterns represent *reradiations* from the line-conductor originating from sparking sources. The number of spikes in a sparking group varies from one to a hundred or more. This number varies with the size of the gap and certain other variables such as voltage across the gap and the impedance to the sparking current at the source location. With a wide gap we have few spikes per group, and with a narrow gap we have many spikes. We can say in general that the spike density is increased as the gap size is decreased. This can readily be demonstrated.

When sparking occurs, we actually have a "train" of sparking groups, as shown in Fig. 11.

When sparking occurs, we actually have a "train" of sparking groups, as shown in Fig. 11-A.

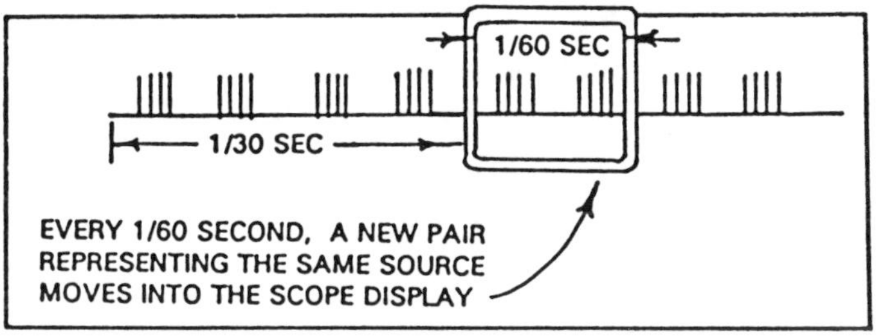

Fig. 11-A. Single Source Radiating a Train of RF-Noise Voltage Groups. This display as detected in your locating system.

Sparking Frequency Spectrum

A belief that has long been held by many is that the wide RF frequency spectrum of RFI from power lines results from the harmonics of 60 Hz. This is simply not true, and the *very wide frequency spectrum* of sparking radiations results *only from the extreme sharpness* of the sparking current pulses. The definitive proof is given for this in Myths 1 and 2 in Chapter 8.

The figure below shows sparking spikes resulting from a DC source, to substantiate the statements above about AC sources. You might think about this and get a full explanation in Chapter 8.

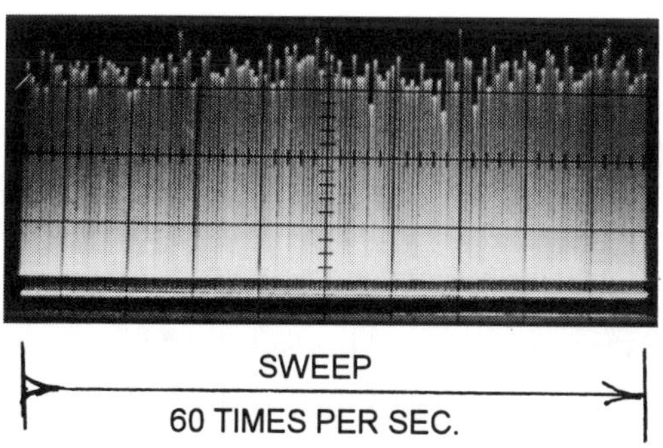

Fig. 11-B. Sharp Noise Spikes from a DC Source with a Horizontal Sweep Rate of 60 Hz. Observed at 76 MHz.

The Power-Line Electric Field

Fig.12 shows how electric lines of force surround an energized power-line conductor. These electric (or electrostatic) lines are present whenever a power line is energized; no load need be present. All we need is voltage on the power-line for the electric field to be present. For sparking occurrence, we are concerned very much with the *presence* of this *electric* field.

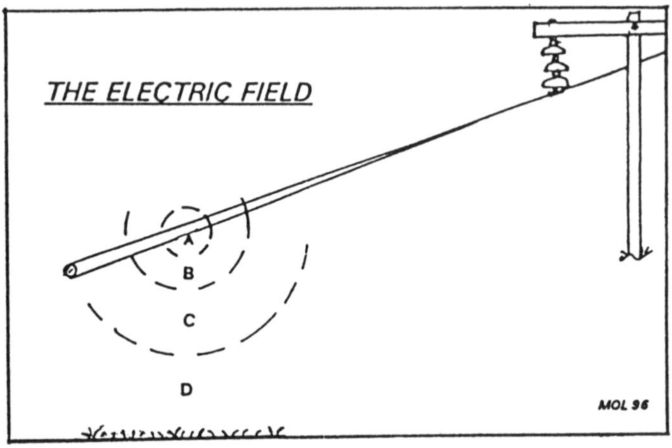

Fig. 12. Electric Stress Lines Between Power-Line Conductor and Ground.

Here, with the conductor energized, a voltage division is present across the total airspace from the conductor to ground. Also, the partial-circle dotted lines represent *equipotential lines* between the conductor and ground. The spaces between them represent equal voltages, and the voltages across spaces ***A, B, C*** and ***D*** are the same.

If we wish, we may designate these lines as "electric field lines," or "stress lines," or "lines of force." If we bring a metal object into this field, it will distort the field lines and assume a voltage induced into it from this field. *Non-conductive objects such as porcelain or polymer insulators will also distort the field to some extend and become charged with a voltage.*

An example of the electric field causing RFI is illustrated in Fig. 13. Here we see a cross-section view of a line-conductor resting in the top of a porcelain pin insulator unit mounted on a wood crossarm. The tie wire holding the conductor in place is not shown. The loose hardware at the bottom end of the pin-bolt of the insulator assembly is charged by the electric field from the line-conductor and is sparking from nut-twasher.

Fig. 13.
RFI Source:
Hardware Sparking
in Electric Field.

The noise source of Fig. 13 is a fairly common one. Sparking is occurring at the loose nut-washer location, and its RFI is induced into and *reradiates* from the line conductor.

The Power-Line Magnetic Field

For radio-frequency noise (RFI) generation, we are *not* concerned with the power-line *magnetic* field. For audio *voice-frequency* interference to telephone systems, however, we are *very much concerned* with magnetic induction from a current-carrying power line and its 60-Hz audio harmonics. Strictly speaking, this is not electromagnetic interference, but is "magnetic induction."

The power-line *magnetic field* is present *only* when there is *current flowing* in the power-line conductors, that is, when there is an electrical load on the line. This "electrical load" usually means the load which results from connected devices drawing current. It may also result from line-fault arcing, usually where current flows from line-conductor to ground.

Sparking, Conductor to Porcelain

It seems logical to assume that small arcs can occur only between metal "electrodes," but this is not so, and we can have sparking between metal and porcelain. A photograph of sparking occurring between a tie-wire and a porcelain insulator is shown in Fig. 1-14. The author believes that this sparking occurs because the electric lines of force in the surface of the insulator have established voltages differing from the voltage on the tie-wire, resulting in spark-overs. This is a transmission-line insulator with 50 kV applied.

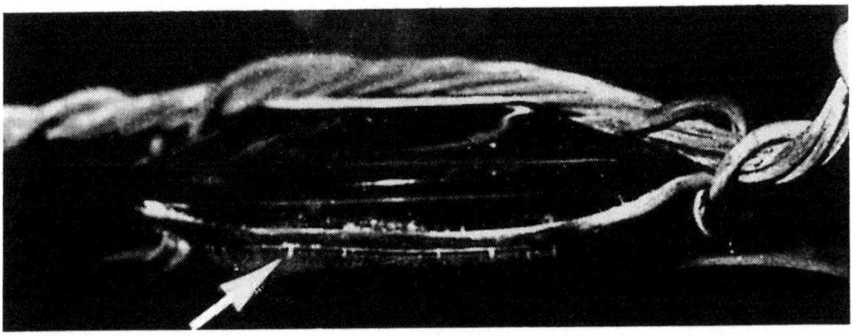

Fig. 14. Sparking, Tie Wire to Porcelain Surface. A transmission-line insulator with 50 kV applied. [Warburton Ref. 3].

Porcelain insulators are often coated in the tie-wire and conductor area grooves with a semi-conductive glaze applied during firing of the porcelain. The glazed area is shown in Fig. 1-15. This coating "equalizes" the electric lines of force on the surface, when the power line is energized. This reduces the sparking and also provides some surface heating to reduce icing.

Fig. 15.
Porcelain Insulator Unit Showing Area of Application of "Semi-conductive" Glaze.

The insulator manufacturers call this glaze a "*semi*-conductive" layer. Actually, it is a "*partially* conductive" layer, or a variation of the carbon resistor. [In solid-state electronics, the term "semi-conductive" has a somewhat different meaning.] This glaze is also applied internally in the pin bolt area, not shown. Such a treated insulator is often called a "noise-free" insulator. On a power line, however, it is not always completely noise-free. For noise-free operation, it is still necessary for the insulator to be tight on its pin, and for its tie wire to be tight on both the insulator and the line-conductor.

A Myth: Most power-line problems are caused by "bad" insulators.

No. *Most* power-line noise problems, considering both distribution and transmission lines, are caused by "bad" connections, that is, arcing/sparking in loose and/or corroded metal to metal connections (which are often in insulator-unit hardware, where the insulators are OK). See other myths in Chapter 8, Sounds, Patterns and Myths.

A Common Metal-to-Metal Example of Sparking.

Metal-to-metal sparking is the most frequent cause of the broad-band RFI concerning us.

An often-observed noise source is metal "trash" on a line-conductor. In Fig. 16, a broken and corroded coat-hanger-wire is resting on a distribution-line conductor. The wire is charging and recharging each half cycle (to a somewhat lower average voltage than the power-line conductor) in step with the sparking voltage pulses, and is sparking over.

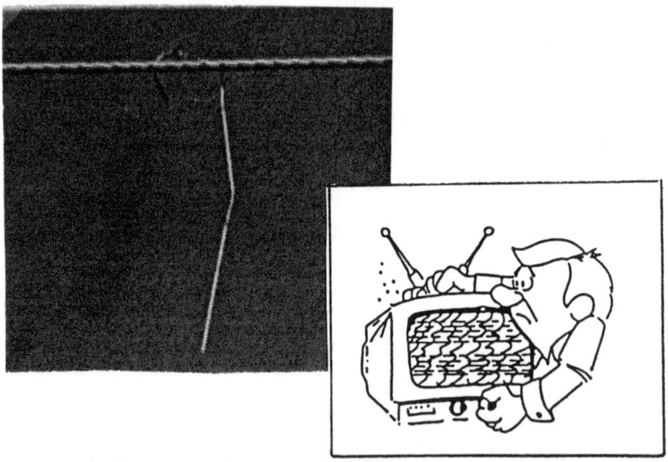

Fig. 16. "Trash on Conductor" Causing Power- Line Noise. As shown in the inset, the party living across the street from this sparking coat hangar could not tune out the resultant TVI!

Relative RFI-Noise Levels in the Locating Spectrum

Fig. 17 gives you a good idea of the relative strength of the VHF/UHF frequencies you will observe, resulting from power-line sparking,

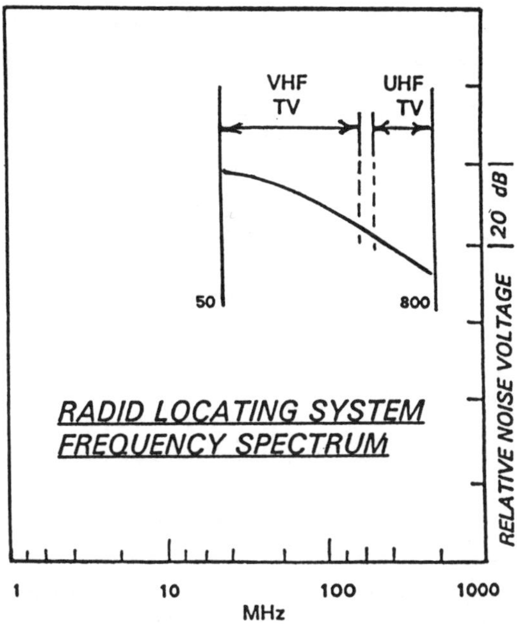

Fig. 17. 50-800 MHz Noise Frequencies vs Voltage Level on a Distribution Line, Measured at 50 ft (15 meters) from the Line. Derived from Pakala-Chartier[4].

On the vertical axis in the figure, an increase of 20 dB means an observed noise voltage increase of 10 times. Lower noise frequencies (at increased levels) are not shown in the graph.

A List of Sparking Characteristics. Table 1-1 on page 22 shows a list of sparking characteristics. Some characteristics shown have not been previously mentioned.

TABLE 1-1. CHARACTERISTICS OF SPARKING.

CHARACTERISTIC	COMMENTS
1. Sparking is visible.	Sparks appear (in the dark) to emit blue-to-white light. On a power line this cannot usually be seen from the ground.
2. Sparking generates little heat.	Attempting to locate sparks with an infra-red detector is not worthwhile.
3. Sparks result from electrical breakdown of air.	This breakdown results in ozone generation which can sometimes be observed by its odor very near the source.
4. Sparks cause audio emission.	This is usually a weak sound not easily detected. Some larger sparks are an exception.
5. Sparks cause super-audio ultrasonic emission.	This can be detected from the ground with sensitive equipment, but only if there is no obstruction between source and observer.
6. Sparking radiates RF.	This radiation takes place over a broad radio frequency spectrum and is readily detectable.
7. This RF is radiated in packets of pulse groups.	The radiation occurs as sparking "spikes," with these spikes occurring in groups; the groups appear in one or (usually) both of the half-cycles of the power-line voltage wave.
8. In an RF receiver, sparking produces various sounds or "noises."	The occurrence of the groups may be intermittent or steady. The "spikes" in a group may be seen from dense to sparse. These factors result in many differences in the sound heard.
9. The spark-generated RF propagates in special manners on power lines.	Propagation is by ***radiation, conduction, induction,*** and ***reradiation.***

SUMMARY

- Power-line RFI is caused by small arcs, called *sparks,* and referred to as *sparking.*.
- Poor electrical connections or loose hardware often results in sparking.
- The main source of power-line RFI on a distribution line is sparking and not corona.
- Sparking sources on a power line account for over 95% of the RFI sources actually found to be on the power line.
- Extremely sharp sparking-current pulses, when present on a power line, generate sharp RFI voltage pulses which cause radiation in a frequency range exceeding 1 to 1000 MHz.
- The use of oscilloscope patterns is a great aid in the understanding of power-line RFI.
- The rise time of an RF sparking spike is about 2 nanoseconds.
- Sparking spikes are seen in groups occurring during each half-cycle.
- There are dramatic differences between a spark (a small arc) and a power arc.
- At a particular location, multiple sources are frequently observed on the oscilloscope.
- Different sources usually have distinctly different scope patterns.
- Sparking can occur between a conductor and porcelain.
- When RFI is a problem, we are concerned about the electric field and not about the magnetic field of a power line.

Chapter Two

THE CORONA SOURCE

A Corona Introduction

A common belief, even among power company people, is that corona is the cause of power-line RFI. Corona on overhead power lines *does* cause interference, but the fact is that, overall, it is responsible for *less than 5%* of power-line sources resulting in complaints. This does not mean that it is not sometimes a very serious source, or cause, of interference. In foul weather it can be serious, as described in *Foul-Weather EHV-Line Interference, Chapter 10* of this handbook.

In driving past a 230-kV (or higher voltage) sub-station, you have possibly heard *corona* in your car's AM radio, with its "crackling" or "frying" sound. This sound is characteristic of fair-weather corona, but to confuse matters, *sparking* heard from your car radio may have the same crackling sound. Here, then, if you turn your radio off, and still hear the crackling with the unaided ear, you know that corona is present. Sparking, of course, may also be present; but its *audio* emission is *usually* very weak. However, with the proper locating equipment, you can readily sort out the sparking from the corona.

A power company person at this substation may inform you that what you are hearing is corona, which is present wherever you have high-voltage lines. This is true, but, even though the corona is present, you may not always hear it, by ear or by radio. Corona occurs in various degrees of sound intensity, depending on certain factors, among these being weather, operating voltage, power-line equipment design, and line conductor age. Corona may sound bad at a substation location, but it is *certainly not* the main cause of power-line RFI.

An Example of Corona

As an example of corona occurring in a 230-kV substation, the open disconnect switch-rods shown in Fig. 2-1 were found to be generating a moderately high level of radio interference at about 1 MHz, as a result of corona pluming at the top of these rods. Operators at this substation complained about the AM noise as they approached this station while driving to and from work (and listening to their favorite, but distant radio stations).

Fig. 2-1. 230-kV Disconnect Switch Rods Shown in Vertical Open Position, after Correction.

To correct the corona in Fig. 2-1 (especially for the sake of research), rounded aluminum caps were applied over the top blunt ends of the open switch rods, as shown in the photo. The resulting rounded ends reduced the electric-field stress at the rod ends and thus reduced the considerable corona activity (and radio noise).

The operators were concerned that the substation was radiating an undue amount of radio noise over a wide area. Actually, that was not the case, and the noise covered only a limited area near the substation.

As a general statement: it is unlikely that any AM radio interference will result beyond 300 feet (about 100 meters) from corona sources in a substation, or from corona sources on a power line operating at 230 kV or below. In a lower voltage switchyard, say 35 kV or below, you may not hear corona at all; here weak corona is present, but it is not significant as an RFI source.

On the other hand, where *sparking* is present, this is another matter, and the radio noise may be observed up to a mile or more--this is as you move along the line-conductor, away from the *sparking* source.

In the foregoing, we have been considering *fair-weather* corona. There are distinct differences between fair-weather and *foul-weather* corona, particularly on extra-high voltage, EHV lines (345 kV and above). On these, corona can cause television interference as well as radio interference in foul weather.

Sparking vs Corona

To compare sparking with corona, we first consider the illustration of sparking in Fig. 2-2.

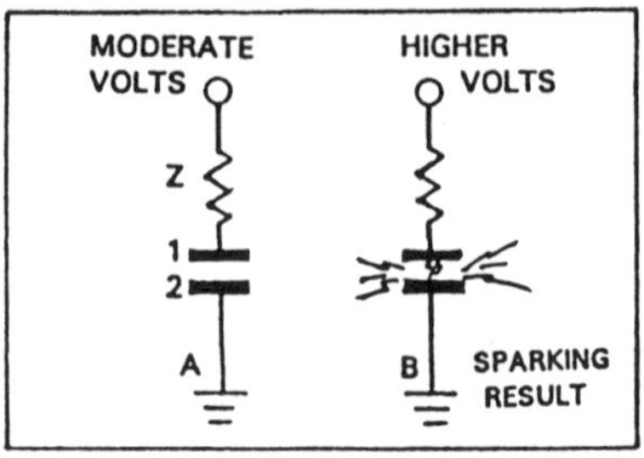

Fig. 2-2.
Sparking in Air Gap. Voltage stress and air breakdown in a metal-to-metal gap. (Z represents the impedance of a power-line sparking circuit.)

Sparking: Referring to the figure above, at A we start with a voltage such that the air in the fixed gap between electrodes 1 and 2 is under moderate stress. As we increase the voltage across this gap beyond a critical level, the air becomes "over-stressed," and, as shown in B, a *sudden and complete air breakdown* occurs, and we have *sparking*.

Corona: Referring to Fig. 2-3, the electric-field lines are shown *parallel* to the conductor rather than *around* the conductor. In (A) we have a protrusion on the power-line conductor which causes the electric-field lines to be "squeezed together," or concentrated, at this point in the electric field. The air is under extra stress at this point, and if we increase the voltage to a critical level, the air becomes over-stressed. *Partial air breakdown* then occurs, resulting in a corona discharge. This breakdown is visible (in the dark) and appears as a brush-like plume. This discharge is not sudden (as compared to sparking) and this is *corona.*

It follows that pluming will take place on even smaller protrusions when we increase the conductor voltage. A protrusion may be a burr on the conductor, a water droplet, a splice compound bead, an ice bead, an insect, wind-deposited plant or soil particles, or other trash. A sharp point on a conductor, then, will be the first location to "go into corona" as voltage is increased, as in (B):

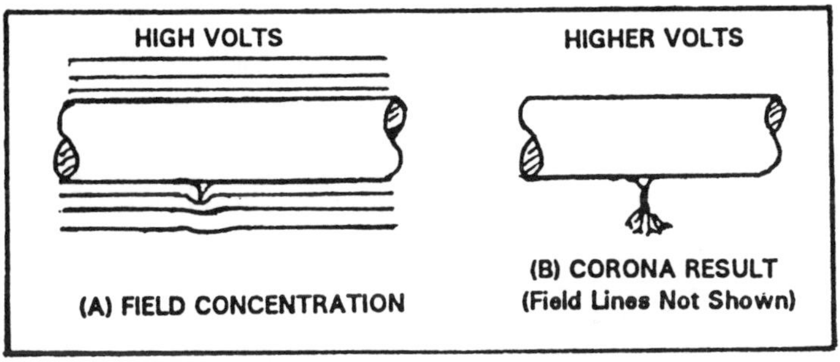

Fig. 2-3. Corona in Metal-to-Air. Voltage stress and partial air breakdown, metal-to-electrically-overstressed air.
(A). Equipotential Lines Around Conductor.
(B). Partial Air Breakdown Resulting from Overstress.

Corona Observations. The discharge corona resulting on EHV lines can often be heard with the unaided ear in either fair or foul weather. In the author's experience, in *foul* weather, a corona plume may extend six inches (15 cm) or more out from the line conductor, and here audio noise (AN) may also result in complaints, in addition to RI and/or TVI complaints.

The corona discharges you hear, and sometimes see, are taking place *only on the positive half-cycles of the power-line voltage wave.*

Both the corona sound and the RFI intensity *increase* as you *decrease* your observing frequency–as you will hear in your radio speaker and see on your oscilloscope screen. Also, in foul weather, corona intensity increases as moisture increases. Further, corona intensity decreases with a decrease in barometric pressure.

The Appearance of Corona. On a dark night, corona appears as a pale blue light. In addition to seeing plumes, you may see a glow, a spray, or streamers. In this book, however, if the corona causes significant RFI, we will consider it to be caused by plumes, or spray plumes.

When you observe corona, for example in a 500-kV switchyard at night, the corona plumes you see may appear to be steady, but are in reality turning on and off 60 times per second. In addition to seeing plumes you may see corona *spray, glow or streamers,* particularly in an EHV substation. In Fig. 2-4 spray plumes are shown.

Some Practical Considerations. If you should receive a complaint of "corona noise" on a transmission or distribution line, you should make sure no sparking sources are present, before you start looking for corona. As one guideline, on distribution lines operated at 25 kV or below, the author knows of no cases where *corona* was the cause of an RFI complaint.

Fig. 2-4. Fair-Weather Corona in 500-kV Switchyard.

Frequencies Observed from Positive Corona Pulses. The upper limit of frequencies radiated by a power-line conductor resulting from corona (or sparking, for that matter) is related to the rise time of the current pulse. A faster rise time means a higher upper frequency limit. The rise time for a corona current pulse is about 40 nanoseconds (as compared to the 2-nanosecond rise time of the sparking pulse). This means there is a great difference in the steepness of these pulses. [Here we are considering the 40-billionths of a second rise time of the corona current pulse to be a slow event!]

> Because of their differences in rise times, the upper RF-radiation frequency limit for *sparking* is over 1000 MHz, whereas for *corona,* it is only about 200 MHz, and corona interference at 200 MHz is very weak.

Negative Corona. You might ask, "Why don't we have corona noise on both positive and negative half-cycles of the 60-Hz voltage, as we do with sparking noise?" Actually we do, but the negative corona is very weak. We are not concerned with this from our practical RFI standpoint. For a theoretical explanation of negative corona, please see the Appendix.

Signal-to-Noise Ratios

In evaluating a complaint of either corona or sparking, it is important to consider the signal-to-noise ratio. Often a favorite AM radio station is either a distant one, perhaps 50 miles (80 kilometers) away, or a weak local station. In quiet areas of minimum interference, very good reception may be obtained, but at 60 ft (about 20 meters) from a moderately strong corona source, reception may be very poor. Here the ratio of desired signal to undesired noise is the determining factor; as you move away from a fair-weather *corona* source this ratio increases rapidly and your reception greatly improves.

Fig. 2-6 shows the effect of corona (or any RF source) as related to the signal-to-noise ratio at a particular location. This figure applies to fair-weather sparking as well as to foul-weather corona interference.

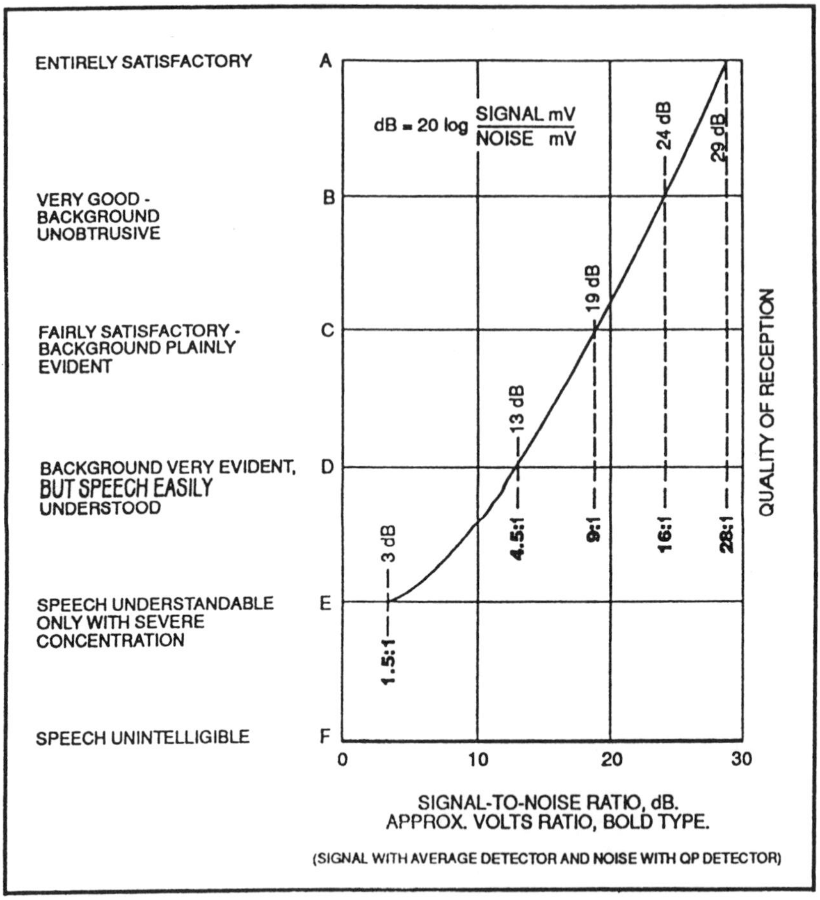

Fig. 2-6. Signal-to-Noise Ratios Required for Quality of AM Radio Reception [Adapted from IEEE Committee Report, Ref. 5].

SUMMARY

- Corona causes less than 5% of power-line RFI-noise complaints.
- Corona *is* a serious cause of RFI on EHV lines in *foul weather.*
- Corona may sometimes sound like sparking and vice versa.
- On distribution lines below about 35 kV, weak corona is present but is not a significant RFI source.
- In fair weather, corona RFI is limited to a few hundred feet from a substation or power lines.
- The rise time of a corona current pulse is much slower than that of a sparking current pulse. With sparking pulses, RFI is more intense and extends to higher RF frequencies.
- Corona intensity increases as moisture increases.
- Corona intensity decreases as barometric pressure decreases.
- Corona radiation is more intense at low frequencies than at high.
- A sharp point on a line conductor or hardware will be the first location on a line to "go into corona" as voltage is increased.
- Corona represents a *partial* breakdown of air as compared to sparking which represents a *complete* breakdown of air.
- Corona is usually visible only in the dark.
- Corona may appear as a faint glow, spray, streamer or plume--any of which (in varying degrees) may cause RFI.
- On an EHV line in foul weather a corona plume may extend 6" or more out from the line conductor.
- In this book, corona plumes are considered the cause of corona RFI, whether it is actually plumes, streamers, spray or glow.
- RFI intensity from corona increases with plume size, as does audio noise.
- It is the *positive* corona that causes corona-RFI.
- Negative corona is not a significant RFI cause.

Chapter Three

NOISE PROPAGATION

Introduction

The locating of power-line RF-noise would be a fairly simple problem if the noise emission at a sparking source were limited to direct radiation. However, this noise travels along or "down" the conductor in both directions–that is, it *propagates* along the power line and *reradiates* from the line-conductor and/or line devices at many places along the line.

Traveling down the line, then, is a bundle of many, many frequencies–these are noise frequencies which have resulted from the very sharp current pulses in the spark-gap of the source. The higher noise-frequencies fade out rapidly as you move away from the source pole, so the sparking-noise level seen on the oscilloscope may be 3 or 4 times as high at 50 MHz than at 200 MHz.

The bundle of noise frequencies being conducted down the line gets smaller and smaller as the higher frequencies bleed off. [Traveling through very thin air or outer space where there is no bending of waves by the atmosphere, there would be no such weakening of the higher frequencies relative to the low.]

Standing waves and/or reflections of various noise frequencies are present along the power line, but we are not concerned with various individual frequencies; rather, we are concerned with the very broad-band noise bundle that causes *reradiation* during its travel.

A Propagation Example

A distribution transformer, as shown in Fig. 3-l, is often replaced because *electrical* noise is heard here on an AM car radio or on a transistor radio tuned to say, 1000 kHz (1 MHz).

This would seem to indicate a noise source on this transformer pole, and the transformer is assumed to be the source. In most cases, the noise heard *does not originate* in the transformer, but has been *propagated in* from another source, and the noise is being *reradiated* at the transformer pole (as well as at other locations on the power pole and the power line.)

Fig. 3-1. The Distribution-Line Pole Transformer.

The use of an AM radio provides only a clue that there are sources out there. It will not usually provide the definite answer that, "Yes, this is a source structure." [*Source structure* in this book means a pole or other structure that contains one or more sparking sources.] *Pinpointing* each of the sources *on a structure* is a separate procedure from locating the source structure. This is described in Chapter 4.

The power company will very often find that the electrical noise they observe on a power line is *reradiation* from a source originating in a home or other building. The company does not want noise emitting from their lines, whatever the cause, and the *usual* practice is to locate and identify the non-power line electrical source.

It would be helpful if you could just take your AM transistor radio, walk up to a power pole and, because of the noise heard, say there is a source on this particular pole. Unfortunately, you cannot do this. The fact is that RFI-noise propagating from a source reradiates at *many* locations on a power line. If the noise appears to be a pole transformer there is *at least a* 90% chance that it is not, and also at least a 50% chance that it is an electrical source in a house or building connected to the power line.

AM-Radio Noise, in this book, generally refers to noise such as that heard on your automobile or other AM-broadcast radio, in the approximate frequency range of 0.5 to 1.7 MHz. The author often refers to sparking noise occurring in the broadcast band as 1 MHz noise. This is because 1.0 MHz is in middle of that band, and the sparking noise covers the whole band.

AM, that is, *amplitude modulation*, is also used for many other services throughout the RF spectrum besides the broadcast band, such as the aeronautical band of 108 through 138 MHz, and the weather band on 162 MHz.

Electrical Interference here means interference caused by electrical arcs, particularly small arcs (or "sparks"). Such sparking causes RFI over a very broad frequency range: beyond 0.5 MHz to 1000 MHz. These sources are found on the power line and off. Some very common off-the-line sources that cause serious noise, with their noise being conducted into and propagated on the power line, are the door-bell transformer, the fish-tank heater, the electric blanket and the electric heating pad..

Non-Arcing Sources such as dimmer switches, touch lamps, photo-cell operated street lamps, stop-and-go street-crossing lamps, security lamps and other electronically controlled devices, when not properly filtered, generate interference in a limited frequency range, usually below 8 MHz.

RFI Propagation Modes on Power Lines

From the source, the noise frequencies generated by a power-line electrical spark propagate or "travel" in one or more of four modes:

(1) via direct radiation at the source, through air or space,

(2) via conduction, along the surface layer of the power-line conductor (this is referred to as traveling *on the conductor, along the conductor* or *down-the-line,* and you may imagine the conductor as acting as an RF waveguide),

(3) via RF induction, where the noise is induced through the air from one power-line phase conductor to another, or to a service-line conductor, that is, RF-noise current in one conductor causes RF-noise current to flow in another conductor, or

(4) via reradiation, where the conducted or induced RF noise reradiates after traveling away from the sparking-source site and encounters a partial blockage (impedance mismatch) such as a pin insulator. This blockage causes reflections in more-or-less a spherical area at the site, which in this book is called reradiation.

In most cases, all four modes of propagation from a single source are occurring at the same time.

An example of the four modes of propagation is shown in Fig. 3-2. Here you have a single sparking source on a distribution line. You can expect to find propagation of the RFI noise by radiation (solid arrow), reradiation (dotted arrow), by conduction (bent arrow), or by induction (circular arrow). This is as you approach the line section of concern while monitoring from a locating vehicle at 75 MHz (or other VHF frequency).

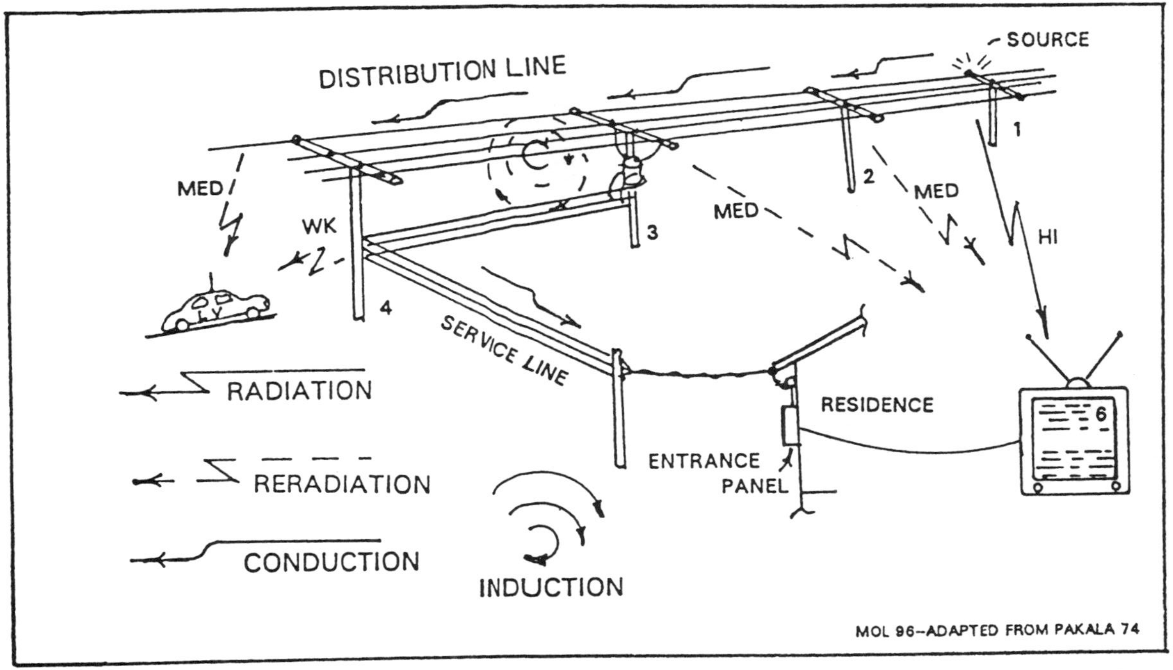

Fig. 3-2. Electrical Noise Propagation Paths from a Single Source on a Power Line. This example is for the TV channel 6 frequency (82 MHz).

In the illustration shown, Fig. 3-2, very little if any noise will arrive at the TV set via *conduction* at any UHF or VHF frequency; however, at lower frequencies, conducted noise could well be detected by an AM-broadcast radio placed near the house wiring.

At VHF frequencies such as at 75 MHz, radiation, reradiation, conduction and induction will take place in a manner typical to that shown, and you could expect to find a medium level of *reradiation* at pole (4), a somewhat higher level at poles (3) and (2), and a high level of *radiation* at pole (1).

The noise monitored near pole (1) would be highest at the furthest phase conductor because the source is on that phase.

On the phase conductor, to which the transformer is connected, higher frequencies, such as TV, will by-pass to ground more readily than will the AM-radio frequencies.

You will find *peaks and valleys* in noise levels as you move along the line and scan the frequencies in the bundle. *Very near an external source,* you will <u>always</u> be able to detect the ultra-high frequencies (UHF), the very high frequencies (VHF), and the medium frequencies (MF).

By "very near" is meant within a few feet of the spark. Observing from the ground, or from your car, you will probably not be able to observe the higher UHF frequencies in this range. However, with proper equipment, you will usually be able to detect, at ground level, noise frequencies from 0.5 MHz through 500 MHz*—if there is a source on the power-line structure you are looking at.

*The 0.5 to 500 MHz range is within the range of the practical locating instruments used by the author.

Internal sparking-device noise. Where we have sources within devices in metal cases, such as in power-line transformers, regulators and breakers, the *higher frequencies* may be shielded by the metal cases, or otherwise filtered out, preventing radiation or propagation from the devices at these higher frequencies. Such internal sparking also occurs in *underground* power cables.

Off-the-Line Sources and Reradiation

RFI from sources in a complainant's house (or his neighbor's house) traveling along the house wiring and then along the service line may be conducted or induced into the power-line phase conductors and then readily reradiate from these conductors. From a power company perspective, this might be considered "reverse propagation."

Corona Propagation This chapter has been concerned mainly with the propagation on the power line of RFI resulting from electrical sparking, and it does not in general apply to the weak corona emissions on distribution lines. These weak emissions will be detected with a sensitive sniffer device, but they are not causing interference and, with a little experience, you will learn to disregard these.

For *practical purposes*, we may consider that power-line *corona* noise occurs only on *transmission* lines and is *observed there mainly* by direct radiation from *corona* sources.

Propagation vs Frequency (and Wavelength)

The propagation along a power line is definitely related to frequency. The lower frequencies, AM radio and amateur-radio lower frequencies travel greater distances down the power line than do VHF and UHF frequencies. The higher frequencies of the broadband noise bundle being propagated are reduced and "wiped out" more and more as we get further from the source. On a distribution line after five or six spans from the source, the only frequencies we observe may be in or near the AM-broadcast band. This is assuming only a single source is present.

Radio waves, of course, (including that of television and all other radio frequencies) have different lengths, as determined by their frequency.

These wavelengths can be visualized with reference to their frequencies by comparing their lengths to some typical distribution-line conductor spans, as shown in Fig. 3-3. [For your reference, the *wavelength* of a radio wave in *meters* is equal to 300 divided by its frequency in MHz.

Ultrasonic Propagation

The emission of ultrasonic waves may be caused by either sparking or corona. For propagation *from* power lines, this "super-sound" *does* require an atmosphere.

Such waves are *not* radio frequencies. This means that they are not a part of the electromagnetic spectrum. However, they may be observed along with RF noise. This ultrasonic emission is present, of course, only at a sparking or corona *source* and not at a *a site of reradiaton*. The ultrasonic component, if present and indicating a source location, will be detected with the proper instruments.

> To best accomplish rapid source locating, which is a prime focus of this handbook, you should accept that power-line RF noise propagates by radiation, conduction, induction, and reradiation (a somewhat unconventional term used by the author).

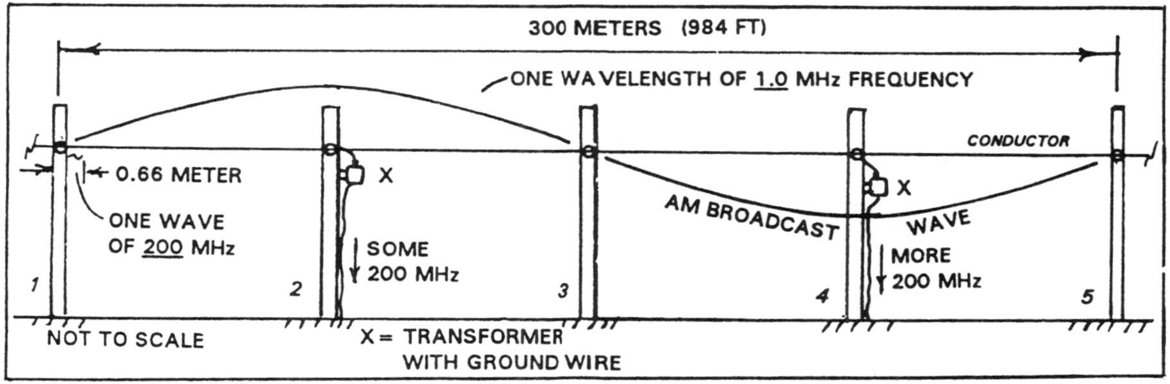

Fig. 3-3. Comparison, 1.0 MHz vs 200-Mhz Wave on Distribution Line.

All frequencies are attenuated (reduced) in strength more rapidly in traveling down *distribution lines than in traveling down transmission lines.* Also on distribution lines, the higher frequencies are attenuated more rapidly than the low. Noise frequencies are arriving from a source to the left of the picture, and some of the high frequencies will be readily by-passed to ground at pole 2, and more at pole 4...As a substitute for more advanced theory, you might think of it like this: The little short waves like 200 MHz can readily squeeze through the pole transformers and go to ground, whereas the big 1-MHz waves are not inclined to do so!

Propagation on Transmission Lines

In general, all frequencies will propagate further along transmission lines than along distribution lines for the following reasons:

1. There are no distribution-line pole transformers on transmission lines that will reduce the level of the noise as it passes each transformer pole, which would bypass some of the noise to ground. (The higher TV frequencies are more easily bypassed to ground than the lower. This bypassing could be via the case of the transformer and its case-to-ground connection.)

2. Transmission lines have longer insulator assemblies (or longer post-type units) supporting the conductors than do distribution lines. These present a higher impedance to the RF noise than the shorter assemblies, preventing the noise from being as readily bypassed to ground, and thus "keeping it on the line."

3. Transmission lines have much longer spans with less opportunity for RF noise being bypassed to ground by devices on the line, as with distribution lines.

As an example of transmission line conductors in one area on a 115 kV line, RF noise at 200 MHz was readily detected at a structure over a mile from the source; here, the source was on the conductor (trash sparking to the conductor). Propagation of noise at 200 MHz for this distance would not be expected on the usual distribution line, with short insulators and transformer poles. As you can see, propagation is related to type of line, type of insulators, number of transformers or other devices on a line, and to frequency.

Underground Propagation.

Propagation of RF noise in underground residential distribution (URD) cables also readily occurs. As with overhead lines, the noise traveling through the cable is reduced much more at higher frequencies than at lower frequencies.

Fig. 3-4 shows a "concentric neutral" URD cable. The outer black vinyl covering has been partially removed to expose the strands of copper wrapped around the white insulation. These copper strands are the concentric-neutral phase conductor.

Fig.3-4. Example of Underground Concentric-Neutral Power Cable.

The white plastic "cross-linked polyethylene" serves as insulation between the stranded copper core and the concentric neutral strands.

The copper strands of the neutral phase do provide some strength and protection, provide some shielding of the electric field surrounding the center conductor, and some, but not complete, shielding of RFI.

A TV cable company would not use such a cable, of course, because it would not be RF-tight and would cause excessive TV signal leakage.

[Starting here, some first-hand stories from the field, told in the first person by the author and others, will be found in various locations throughout the book; however, most of the stories are grouped together in the collection of Chapter 9].

STORY 1

AN UNDERGROUND SOURCE & 2-MILES of RERADIATION

(by mol)

An interesting search for noise on *amateur and AM frequencies* started on a noisy two-mile section of 12.5 kV overhead line, and ended on an underground cable. Along two miles of the power line, which was *reradiating* noise at AM and ham radio frequencies, the highest frequency that could be observed was near 6 MHz. This observation was close to a condominium, in a suburb of Seattle, served with a 12.5 kV URD (underground residential distribution) system, and at the URD riser pole similar to that shown below.

Fig. 3-5. A 3-Phase Riser Pole. This is often called a Dip Pole by the people up in the air (the line crew).

A power-line crew was called out and the various vaults and pad-mounts in the system, starting at the riser-pole vault, were individually opened. As this was done, the noise was traced by identifying the noisy disconnects on the underground cable in each vault with the RE 247 hotstick sniffer, as in Fig. 3-6. This noise was traced from one vault or pad-mount cabinet to another, and at each location was found to be on the same phase-conductor.

After five covers had been opened, an exceptionally noisy elbow-disconnect was located in a vault at about 400 ft. of cable from the riser pole. The sniffer is shown pinpointing the source.

Fig. ***3-6.*** Pinpointing the Source in Transformer Vault with RE 247 Hotstick Line-Sniffer.

SAFETY FIRST! Leave it to the certified lineman or serviceman. Touching any parts in vault could be fatal!

Fig. 3-7. Elbow Disconnect.

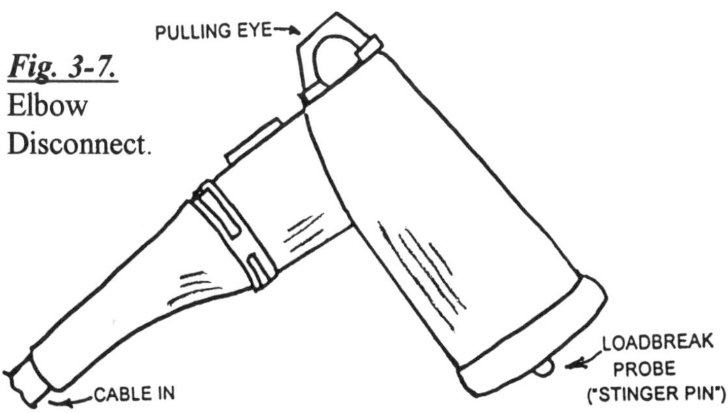

After de-energizing the URD system, the crewman found a loose center-conductor stinger-pin in the elbow-disconnect.

He found he could tighten this stinger by one-and-a-half turns; this being done, he replaced the disconnect into its bushing socket, re-energized the cable, AND two miles of radio noise was cleared up! This was much to the satisfaction of the complainant, who had advised the FCC; they had advised the president of the power company, who had advised his crews to *fix it!*

A Question: "How was it that we could not detect the noise above 6 MHz, observing from the ground, but could trace it via the vaults using the short antenna rod of the hotstick sniffer?" The answer is that, if we had sniffed the URD cable terminal bushings at the top of the riser (which we did not), we would likely have detected *weak* 110-MHz noise close to the terminals. This weak noise *reradiating* here was not detected at ground level at VHF frequencies but was detected there at the low frequency of 6 MHz.

Frequencies of Concern.

The classification of frequencies in the radio spectrum are shown in Table 3-1. We are most concerned (from an RFI propagation and locating standpoint) with the frequencies as shown bracketed by A.

For the practical purposes of this handbook, the power-line/electrical RFI-noise locating to be described is done between 0.5 MHz and 800 MHz. Usually, the higher frequencies in this range are the most effective for direction-finding of a source and for confirming its location from the ground, as described in the next chapter.

Table 3-1. Radio (and Noise) Frequency Classification

FREQUENCY	CLASSIFICATION
3 to 30 kHz	Very-low frequency (VLF)
30 to 300 kHz	Low frequency (LF)
300 to 3000 kHz ←	Medium frequency (MF)
3 to 30 MHz A	High frequency (HF)
30 to 300 MHz	Very high frequency (VHF)
300 to 3000 MHz ←	Ultra-high frequency (UHF)
3 to 30 GHz	Super-high frequency (SHF)
30 to 300 GHz	Extremely high frequency (EHF)

STORY 2

IS THAT CB-NOISE CORONA?

(by mol)

Shortly after Citizens Band (CB) Radio became popular, I was called to check out a complaint along a transmission line corridor in the wheat fields of Eastern Washington (Fig. 3-8). This was from a CB operator living near the corridor in which run two 230-kV lines and one 500-kV. Could it be corona?

Even though there were threatening clouds as a backdrop, it rains very little in the area, and "foul-weather corona interference" was not likely the cause. But at 27 MHz (the CB frequency), fair weather corona interference was possible...I thought. At this time, many people thought corona to be *the* cause of power-line interference. However, my oscilloscope said this was a single sparking source.

Fortunately, the maintenance trail in the corridor made it easy to drive along the lines. Using an early version of my rapid locating system, I first located the general area of the source from observations made from within my vehicle.

Then, with my hand-held locator, I located the source as being on a 230-kV line-conductor and observed with my binoculars a length of baling wire hanging on the conductor. The source was found! It obviously was wire trash sparking to the conductor and not a case of corona.

The line maintenance crews readily removed the baling wire, and the radio interference was eliminated, permitting the CB'er who lived about two power-line spans away (in the distance in this photo) to happily continue his CB operating.

I later found that "fair weather" corona noise would not be detected at CB frequencies (that is, at or near 27 MHz). However, it can be readily detected on an AM broadcast receiver from a 500-kV line in fair weather.

In foul weather, however, at 500 kV, corona noise is observed at VHF frequencies. See Chapter 10 for foul weather TV interference.

> Sparking interference is the dominant RFI found propagating on power lines and resulting in complaints to a power company, and in general, all frequencies will propagate further on transmission lines than on distribution lines.

Fig. 3-8. Telescopic View of High-Level RFI Source on a Transmission Line of the Bonneville Power Administration.

SUMMARY

- RFI (sparking noise) originating on the power line or in a home is *propagated on* and is *reradiated from* the power line.
- Noise *appearing* to originate at a pole transformer *usually* is *reradiation* originating from sparking at a different location.
- An AM broadcast radio cannot conclusively locate a power-line pole as being an RFI-source pole.
- *Electrical* interference equates with RFI caused by *sparks* (or arcs) *on or off* a power line.
- RFI propagation modes are radiation, conduction, induction, and reradiation.
- From a single sparking source the four modes are almost always present and are all occurring at the same time.
- Sparking RFI-noise can readily propagate in underground cables.
- Sparking interference is the dominant RFI found propagating on power lines and resulting in complaints to a power company.
- Sparking noise from a power-line source starts its travels along the power line in a bundle of many, many noise frequencies from below 1 MHz to above 1000 MHz.
- Distribution transformers are often blamed, but seldom cause, power-line noise.
- AM (amplitude modulation) is used in AM-broadcast, in aircraft band frequencies, and in other services.
- In general, all frequencies will propagate further on transmission lines than on distribution lines.
- Underground sources can be traced through a system with a VHF sniffer (such as a 110-MHz sniffer).

Chapter Four

RAPID NOISE-SOURCE LOCATING

Introduction

In Chapter 3, *Noise Propagation,* you were introduced to noise-locating in two stories where each illustrated propagation of a single-noise source. The following discussion, however, will concern source-locating with multiples sources present.

Here we illustrate that you can see what's really going on out there in the electromagnetic environment with the use of an oscilloscope–and you can use what you see to guide you in your rapid locating of power-line interference sources

You will have an instant view of the number of power-line interference (noise) sources present, their relative intensity, and with your rotatable antenna, determine their direction. You can then trace the highest level, and then the lower level noises to their sources. This *cannot* be done by listening to the noise(s) on an AM radio, or an amateur radio receiver.

An "instant view"of two sources seen on a cathode-ray (CRT) oscilloscope is shown below.

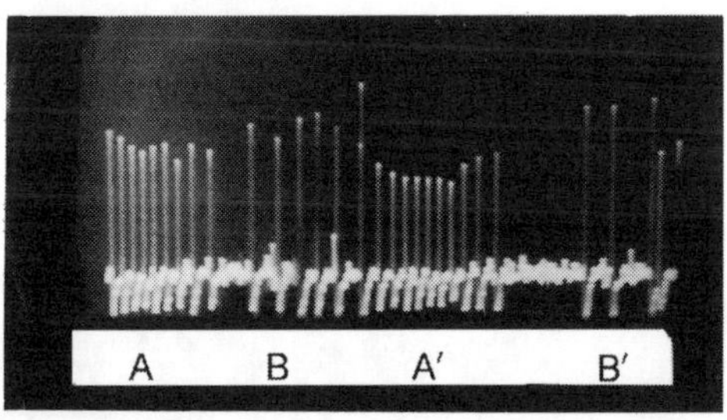

Fig. 4-1. Two Power-Line Sources Seen on Oscilloscope.

The higher pair (B & B′) is from one source, and the many-spike pair (A & A′) is from a second and probably more-distant source.

In your vehicle, in front of the customer's home and parked near the power line, you often see two or more sources. The highest level spike-group of *two* spike-groups seen may represent the source which is the whole cause of the complaint. However, if near the customers home you observe *two* sources, you should locate and correct the *two* sources.

Using the method of this chapter, you will be able to locate those power-line sources in less than half the usual time,

Other methods will also be described in Chapter 6 to suit the equipment you may have on hand. These are not as efficient but are useful if you have only an occasional interference problem.

Equipment Requirements

For the locating method, certain equipment is required, and some examples of such locating equipment are shown.

To start, an excellent antenna for locating the *general area* of the sparking source is shown in the figure below. This is the modified Channel Master 3619. [See the Appendix for details].

Fig. 4-2. Rotatable Antenna on Vehicle Roof. The mast is raised, lowered, and rotated from within the vehicle.

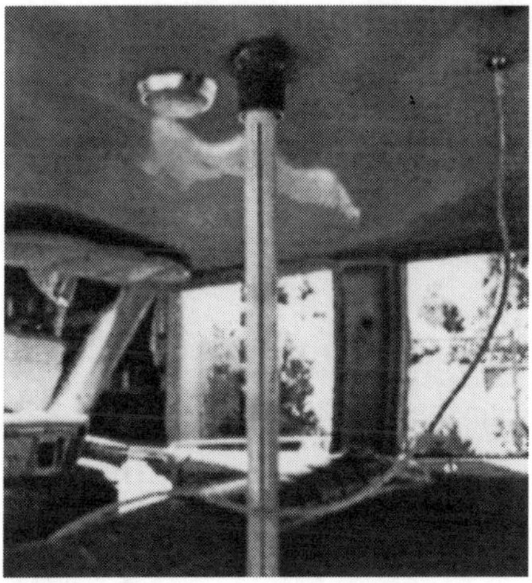

Fig. 4-3. Interior View of Rotatable Mast.

An inter-connection equipment diagram is shown below:

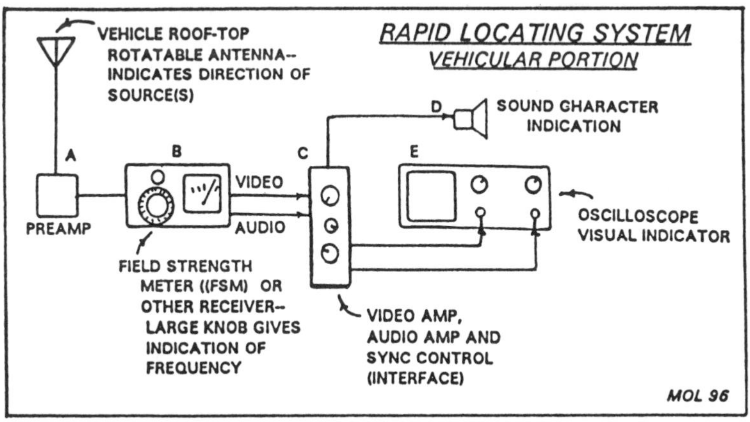

Fig. 4-4. Rapid Locating System Inter-Connections.

[Note: with Radar Engineers vehicular locating system, the items A, B and D are an integral part of their model 239, with item C being a special interface].

Other instruments that may be used in the above are shown on the following pages.

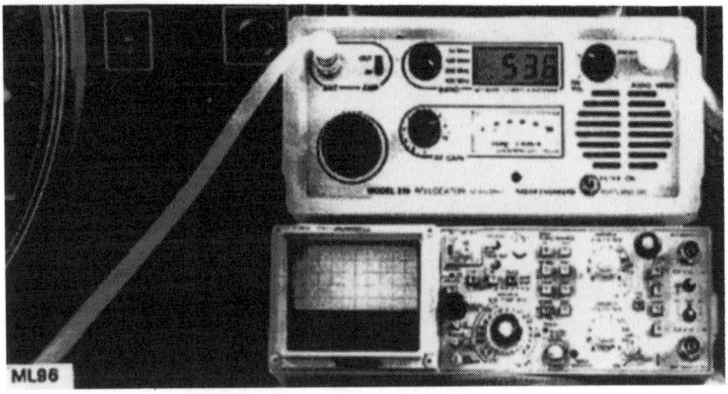

Fig. 4-5. Equipment on Front Seat of Sedan. The RE 239 Locator with the Leader Model 315A scope is an excellent combination.

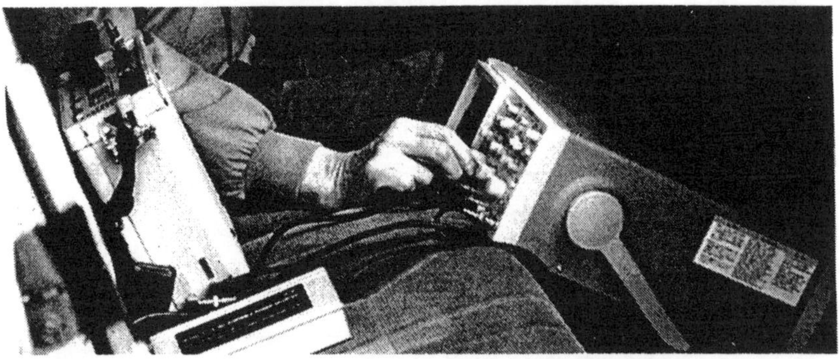

Fig. 4-6. RE 239 Locator, RS 22-145 Converter, & the Kenwood high-intensity oscilloscope, CS-4135–also a very good setup.

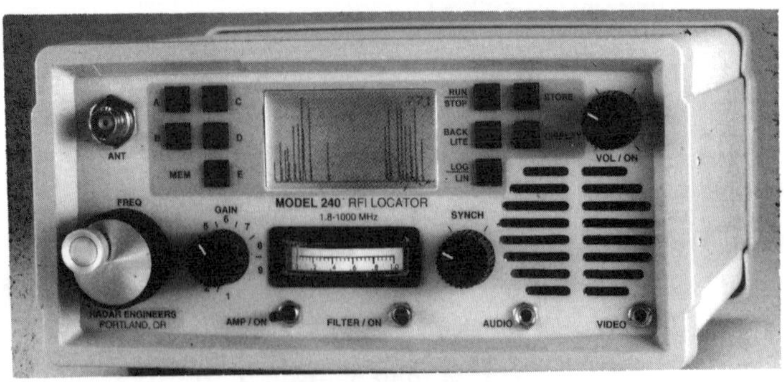

Fig. 4-7. The RE Model 240 Locator.

The Model 240 combines the locator receiver with an LCD scope display in one instrument. An excellent feature of the Model 240 is that it can display and store a pattern seen at the complaining party's location, and compare it with the pattern seen at a suspected structure.

Its LCD (liquid crystal display) is adequate for much of your locating, but it is not as sharp and bright as the cathode-ray display of the scopes of Fig. 4-5 and Fig. 4-6. *For research purposes,* if more than two sources are present, it does not have the discrimination desired.

Oscilloscope Noise Patterns. With your equipment properly adjusted, you will see noise patterns resulting from power-line or other electrical sparking sources. You will see these patterns in various shapes and sizes, and you *can* keep track of them with your oscilloscope.

The Rapid Locating Method

The method described here is intended to permit you to:

(1) ***Rapidly locate the general area*** of a power-line RFI source, that is, the few poles(or structures) containing an interfering source.

(2) ***Identify*** the *source structure.*

(3) ***Pinpoint*** *the source on the pole.*

If non-power line sources are present, to:

(4) ***Locate the non-power line source(s)*** in homes or buildings which are causing reradiation from the power line.

(*1*) ***Locating the General Area of the Source***

A Locating Narative. In the field, your first step in resolving a noise complaint is to visit the customer and see if the interference is on, at a time of day reported to be on. In the example following, television interference is the complaint.

At the customers's home, you see that he *does* have the TVI he had complained about. You recognize from the TV pattern, with its two bands of "dot-dashes" drifting up across the screen, as seen in Fig. 4-8, that it is sparking noise. And you also know that it may or may not be power-line noise.

Fig.4-8. TVI at Customer's Residence.

Your customer says that *this is it*, and says the noise is on the lower channels, 2, 3, & 6 only. It is not on (the much higher-frequency channels) 8, 10 & 12. These are the six channels he receives off-the-air. [See table 4-1 for channel frequencies].

You deduce that the noise is not originating in his home, because if it were, you would also see it on 8, 10 &12. [The lower VHF channels 2, 3, & 6 can travel in on the line from a considerable distance–whereas the higher VHF-frequency 8, 10, & 12 can travel only a short distance. These higher frequencies have been "filtered out" from the noise-bundle along the way, and do not arrive at the residence to interfere with 8, 10, & 12].

You then go out to your vehicle to confirm that there *is* noise reradiating from the power line

In your vehicle you can check the proper operation of your equipment by tuning to a fairly-strong TV channel. A television-video pattern similar to that shown in Fig.4-9-A will be seen. You may see some interference overlaying it.

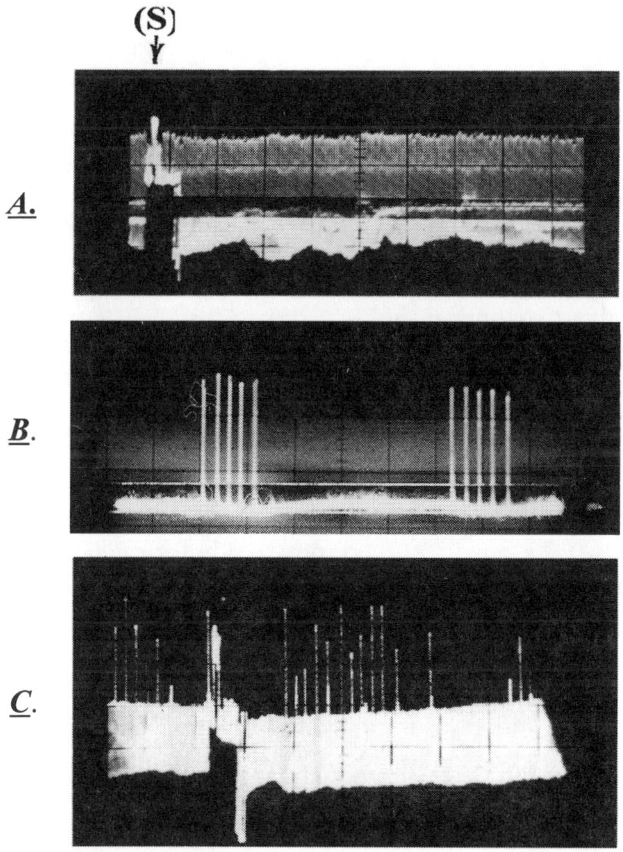

Fig. 4-9. Noise and Television Patterns at 60-Hz Sweep Rate.

A. TV-video pattern. Sync pulse (S) at the left in pattern
B. Noise *spike-pattern-pair* of a single sparking source.
C. Sparking-noise pattern interfering with TV pattern.

[The repetition rate of the vertical sync pulse of the video carrier is a nominal 60-Hz (but precisely 59.94 Hz). If you tune to this with your RF receiver or your FSM, you can hear it as a smooth hum. Early in my locating experience I mistook this "60-Hz" hum for AC interference. So, do not confuse this with the sound of power-line or electrical sparking, which sometimes has a 60-Hz humming/buzzing character].

If you can see the pattern of A, or B, or C, of Fig 4-9,and by rotating your rooftop antenna you change the amplitude of the scope pattern, you know that your antenna is working, and that your equipment is *on*, connected, and working properly.

Then, near the transformer pole at the customer's home you tune your equipment to channel 6, and at that frequency see a sparking pattern on your scope. You then rotate your antenna, and if you note a higher level to the east you proceed to drive east along the distribution line.

You should stop at every few poles to see if you can pick up the same pattern at channel 8 (a much higher frequency). At about 4 poles away you do observe noise on this channel, and after a few more poles you observe the noise on channel 12. You are now in the immediate vicinity of the source and have located the *source-area*. You are now ready for the next step.

(2) Identifying the Source Structure.

In the source area, observing from your vehicle, you probably cannot tell for certain which of three poles has the highest level. You then go much higher in frequency by getting out of your vehicle with your hand-held antenna, and a UHF locator-receiver tuned to 500 MHz. With these you search for the pole with the highest noise level, as shown in Fig. 4-10.

Fig. 4-10. UHF Locator, RE Model 239. (This locator also operates at VHF, down to 50 MHz). The RE Model 240 may also be used in this manner.

Where spacing between poles is good, say 100 ft or more, the author most often uses the one-piece RE Model M-330 UHF Locator for confirming a structure as a source. This is for convenience and speeding up the job, and also because this instrument is easy to demonstrate to new people in the field. These people find the M-330 is a pleasure to use, as does the author, because of its one-piece, light-weight configuration. The M-330 is shown in Fig. 4-11.

Fig. 4-11. The RE Model M-330 UHF Locator.

Because of the very high sensitivity of the M-330 you may experience the following: Standing on the road beside a power line, and fairly close to one of the poles, you might point towards the power line but possibly pick up a noise source a half-mile away, located on another power-line or in a substation or a switchyard–to which you may be also pointing. You should then reduce the gain of your M-330 for your close-up locating.

After locating the source pole from out of your vehicle with your RE 239 or M-330 locator, your have identified the source structure, and find yourself possibly 6 or 7 poles away from the customer's residence. *You are now ready for the next step.*

(3) Pinpointing the Source.

For pinpointing (and incidental correction) of a source on a structure or in a vault (as was shown in Fig. 3-6), you should call for help from a power-company line crew.

If the source in on the overhead, you might thoroughly scan the structure with your UHF locator, to try to identify a small area of noise. You do this scanning using different orientations of the antenna elements–holding the antenna with its elements vertical or horizontal, or in-between, seeking the highest noise level.

If you have an RE Model 250 Parabolic Ultrasonic Locator, you might first scan the structure carefully from ground level. If the parabolic dish can get a direct view of the sparking source, you will be able to detect its location. The Model 250 is shown in Fig. 4-14, page 4-14.

After locating from the ground level by either of the methods above, the precise source location should be confirmed by the lineman using a hotstick line sniffer, as shown in Fig. 4-12. For example, you might have found such a location to be on the left side of a crossarm, and this would be good place for the lineman to start.

The visual inspection. You should not expect that you can tell by a visual inspection of a structure "that everything up there is OK," and that this structure is *not* the cause of the noise complaint received by the company. This is sometimes reported by a line crew not aware that even though everything appears to be tight and ship-shape, there could well be a hidden source, such as a metal pin-to-thimble-insert in a pin insulator. Before making such a report, the crew should thoroughly scan the structure with the RE 247 sniffer with its antenna-rod fully extended.

Fig. 4-12. RF/Ultrasonic Hotstick Line Sniffer, RE Model 247.

A. Checking with the ultrasonic probe for hardware noise.

B. RE 247 showing ultrasonic probe-tube & sensitivity-adjusting antenna-rod on top and hanger-loop on its back.

The pinpointing is done by starting with the antenna rod fully extended to scan for an area containing the source. This 110-MHz RF rod is then pushed in with small steps to decrease its sensitivity and describe smaller-and-smaller areas of noise.

The small area can be further reduced by using the ultrasonic 40,000-Hz probe tube to precisely pinpoint the source. Here the sparking is found to be sparking between the nut and washer of the center bolt to which the probe is pointing in the figure.

A misconception. On one occasion the author was accompanied in his vehicle by a technician on a noise-locating search. The oscilloscope was on and displaying a particular pattern from the vehicular-locating equipment. It looked similar to ***Figure 4-13*** ⟶

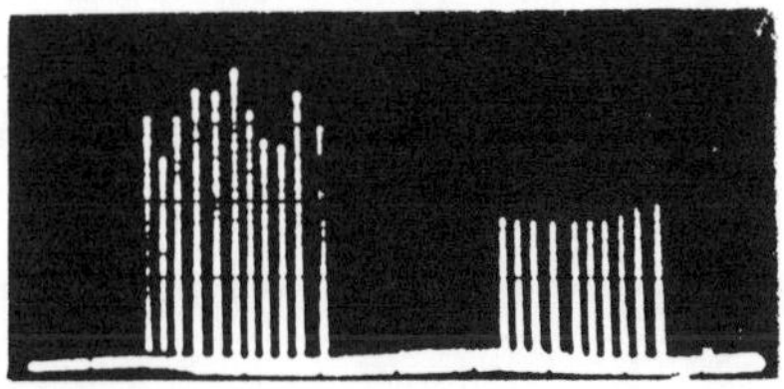

The technician looked at this and remarked that "It looks like we have two sources here."

It had never occurred to me that anyone would think this, but then I realized that a person logically might assume that a display with two different patterns on the screen meant two different sources.

In truth, with a horizontal sweep-rate of 60 Hz, a pair represents a single sparking source. This with the first group of the pair occurring on the positive half-cycle of the 60-Hz voltage, and the second group on the negative. We have rectification here, and the second group is partially rectified, as discussed in Chapter 1, page 13.

It is important to realize that we are looking for a pair of patterns to identify a particular sparking source, and that this *pair of patterns* is repeating itself every 1/60th of a second.

The Parabolic Ultrasonic Locator is shown in Fig. 4-14. As suggested by the author and others, a parabolic ultrasonic locator specifically designed for pinpointing sparking locations on a power line, and not intended to accomplish many other tasks, is now available. This is the RE Model 250, and the author finds it to be *at least twice* as sensitive for power-line-sparking pinpointing from ground level as any other ultrasonic instrument on the market.

This *one-piece* locator is a *light-weight* parabolic-dish-plus-amplifier assembly. It is most often used *without the necessity of headphones*. These features make it a very useful instrument. [Certain minor inconveniences such as gain-control placement, and the difficulty of battery replacement in the field prevent the author from giving it a "fantastic" rating. Hopefully such a rating will come soon].

Fig. 4-14. The RE Model 250 Parabolic-Ultrasonic Locator.

This Model 250 locator is invaluable if your company requires that you give them the exact location of a source, before they send out a crew to correct it.

The 1- MHz sniffer. On occasion you may have need of a 1-MHz sniffer, such as shown in Fig. 4-15 If, for example, you have a complaint of AM or amateur radio noise (with no accompanying TV interference), you *migh*t have a noisy pole transformer, as the customer says. You can prove or disprove this with a 1-MHz sniffer. Construction details for such a sniffer are found in Chapter 7.

Shop tests and experience have shown that "bad transformers" do not radiate noise above 8 MHz. Thus the RE Model 247, which operates at 110 MHz would not detect them–and you need the *1-MHz* sniffer.

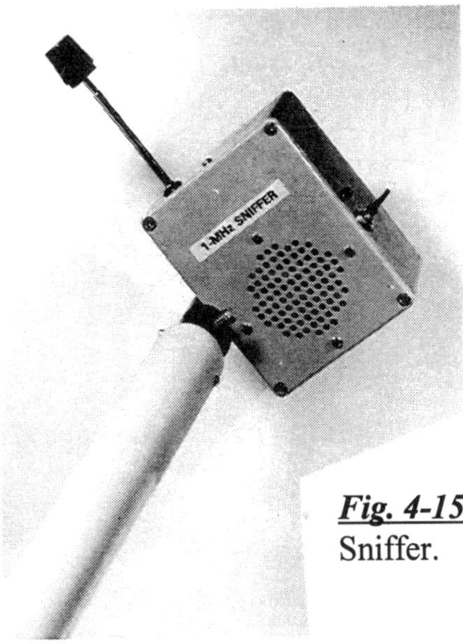

Fig. 4-15. The 1-MHz Sniffer.

Before calling a line crew to check a transformer, be sure to check for noise sources at all the houses connected to this transformer. For this, use the 1-MHz sniffer and hold it near the house entrance meter with the antenna rod fully extended.

Shorten the rod to where the noise can be just barely heard. Leave it at this extension and compare the noise level found with other connected houses. If none of the houses have any (or very little) noise at the entrance meter, you have a bad pole transformer (most likely) and you should call for a power-line crew for help Then have the lineman check the primary bushings of the energized transformer with this 1-MHz hotstick sniffer.

He should use a minimum of rod extension from the top of the sniffer, say 1/2" for low pickup sensitivity. If the transformer is bad, he will hear the sparking noise at these bushings.

An accepted theory is that air-bubbles in the transformer oil provide sites for sparking, and the noise.

Theory aside, the noise problem can be corrected by de-energizing the transformer, applying a vacuum hose to the otherwise tightly-sealed metal transformer case, removing the air-bubbles and then restoring the air pressure to normal. The metal case may then be sealed and the transformer energized.

On one rare occasion, the author located 15 bad transformers in a section of line, using his 1-MHz sniffer. This was on a new line, where a batch of bad new transformers were installed.

Underground checks. The l-MHz sniffer is also useful in checking noise suspected in an underground residential distribution system. If noise is detected at the riser-pole terminals at 1-MHz, but very weakly (if at all) at 110-MHz using the RE Model 247, the noise is coming from the underground.

Antenna rod vs ultrasonic probe. Note in the above paragraphs that the author specifies using the sniffer *antenna-rod* for the scanning.

I received a report from one line crew indicating that they *never use the antenna rod of the 247,* and that just the ultrasonic probe works fine for them.To me this was somewhat disturbing.

With this practice they would miss many sources. This is because the *ultrasonic* will pick up a noise source only with a direct view of the source. If the source is blocked it won't do the job. Using the RF-rod with full extension, you will pick up a sparking source even if it is on the other side of a crossarm, for example. After a small area is thus detected, the exact source can then be confirmed by a short antenna-rod extension of about 1/4" and in most cases, with the ultrasonic probe of the hotstick sniffer.

Power-line noise and noise from the thermostats of fish-tank heaters, of doorbell transformers, and of other devices have similar looking oscilloscope patterns. All of these represent "electrical noise."

Intermittent sources are often time-consuming, but these can be efficiently located by concentrating on them during favorable weather conditions, during a time of day reported to be on, *and* using an oscilloscope in your vehicle as follows:

Each time the noise is on for a short but significant time, you will quickly rotate your antenna for peak noise on the scope and let this procedure lead you to the source (or to a reradiating structure). This may include several 10-minute stops, for example, but the waits are worthwhile.

Your vehicular-locating equipment with its oscilloscope and rotatable antenna will bring you close to a home radiating electrical noise, or to within a pole or two of a power line sparking noise-source.

A composite story based on actual noise-locating experiences, with oscilloscope patterns seen along the way, is shown in Fig. 4-16.

Chronology of a Power-Line Noise Search.

See Figure 18. After receiving a complaint of TVI *on channel 2* from a CATV company Ted, the power-company investigator, observes the following:

A. From his vehicle at the cable company antenna head-end site, he sees two pairs of noise-patterns at *channel 6* on his oscilloscope, indicating two sources.

B. Traveling south, sees the same source patterns at B except source-patterns of #1 source are slightly higher.

C. He sees the same two sources patterns.

D. Ted *tunes up to channel 7* and sees only one weak #1 source-pattern pair. This indicates that the number #2 source-noise, not seen at channel 7, is coming in on tap-line C-J-K from the east.

E. #l source-patterns are now seen with a higher level on his scope.

F. *He tunes up to channel 13* and gets weak level on the #1 patterns. He finds moderate #1 noise level at poles G and H.

G. He locates #1 source pole at G by comparing G & H with hand-held 330-MHz locator & calls for the power-line crew for repairs.

After repairs are made by tightening up hardware, Ted goes back to C to check out source #2. Highest level is found at J but source is found to be in the nearby residence. He installs his plug-in "test-filter" on a fish-tank heater cord, and the #2 source is corrected. The home owner is pleased and will buy a new filter, a Radio Shack #15-1111, for Ted.

Ted goes back to head-end site and finds reception noise-free on all channels, and the cable company people are all smiles.

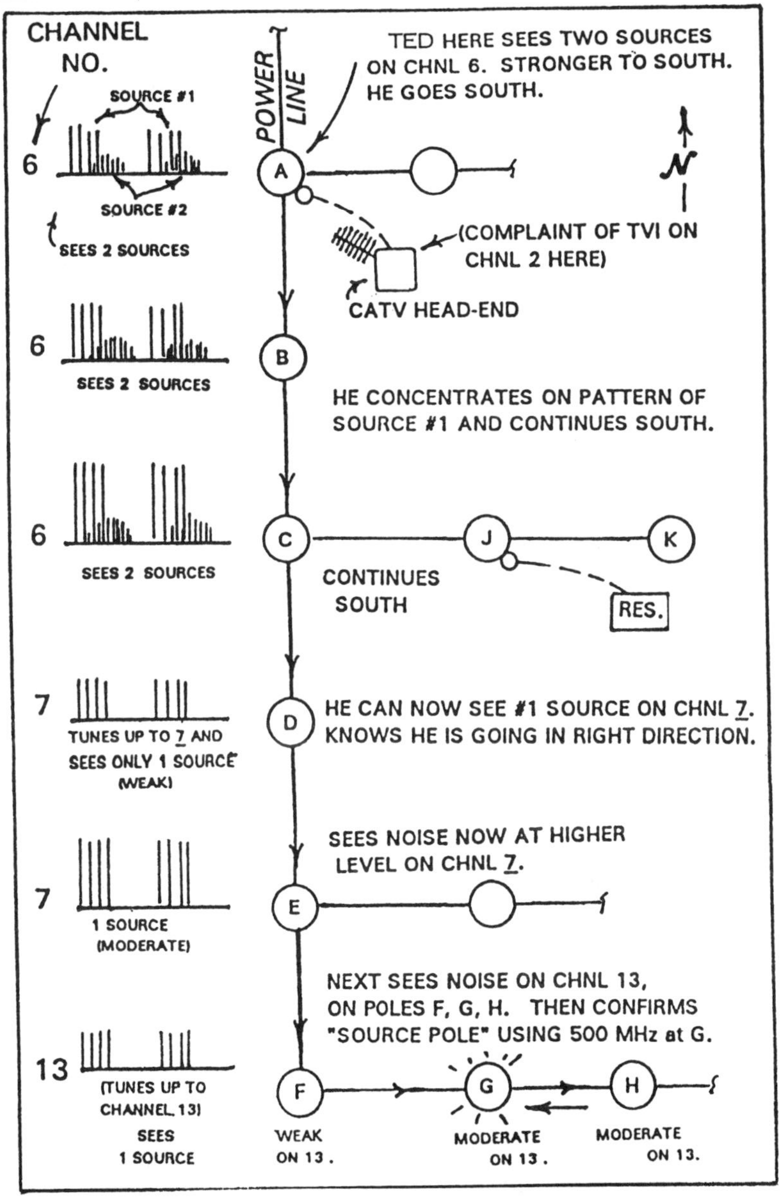

Fig. 4-16. Locating Power Line Noise Sources, an Example.

TABLE 4-1
OFF-THE-AIR TV ASSIGNMENTS by FCC

NOTE:
FOR VIDEO CARRIER
ADD 1.25 MHz TO FREQUENCY
AT LEFT OF CHANNEL NUMBER.

FOR EXAMPLE: VIDEO CARRIER
FOR CHANNEL 3 IS 61.25 MHz.

FREQ. MHz.	CHANNEL.	
54	2	LOW VHF
60	3	LOW VHF
66	4	LOW VHF
GAP OF 10 MHz		
76	5	LOW VHF
82	6	LOW VHF
GAP OF 92 MHz		
174	7	HIGH VHF
180	8	HIGH VHF
186	9	HIGH VHF
192	10	HIGH VHF
198	11	HIGH VHF
204	12	HIGH VHF
210	13	HIGH VHF
GAP OF 260 MHz		
470	14	UHF
476	15	UHF
482	16	UHF
488	17	UHF
494	18	UHF
500	19	UHF
506	20	
512	21	
518	22	
524	23	
530	24	
536	25	
542	26	

FREQ. MHz.	CHANNEL.	FREQ. MHz.	CHANNEL.
548	27	680	49
554	28	686	50
560	29	692	51
566	30	698	52
572	31	704	53
584	32	710	54
590	33	716	55
596	35	722	56
602	36	728	57
608	37	734	58
614	38	740	59
620	39	746	60
626	40	752	61
632	41	758	62
638	42	764	63
644	43 (UHF)	770	64
650	44 (UHF)	776	65
656	45 (UHF)	782	66
662	46 (UHF)	788	67
668	47 (UHF)	794	68
674	48 (UHF)	800	69

Above does not include off-the-air satellite stations.

STORY 3

SOURCES NEAR THE HAM SHACK

(by mol)

In this complaint a radio amateur operator said that he had "bad" noise on 60 to 80 meters. Besides calling the power company, he had also called the FCC. The FCC had been out to his home and advised him that it was power-line interference. Since the FCC was involved, the power company was especially anxious to correct the problem (if it *was* a power-line problem).

The ham operator stated that the noise would usually become severe about 3 in the afternoon. The weather had been fair for some time, with a lot of sun in the afternoons, and this could be a factor.

I arrived in the noisy area a little before 3, and before stopping at the radio amateur's house I made a preliminary short trip around the neighborhood. I observed that there were at least two sources within two blocks of the radio amateur's residence.

When I drove up to his house, the operator, Ernie, was waiting for me. He told me that he had a high noise level on his ham radio right now, and it was reading about an S-9 (on his receiver's signal-level mater). To the ham radio operator, this would represent a very high level of noise.

I told Ernie that I expected that I could get the noise down to an S-3, and he said he would be happy with that.

I could make this statement, knowing that there was not only a high-level noise source nearby, but another one not quite so high. [These were undoubtedly contributors to his problem. Eliminating these would bring the noise level at his antennas way down, and if it wasn't down to an S-3, I would find and have corrected other further-out sources. S-meter readings are only relative and will vary from receiver to receiver. [It is good to say that the readings, before and after, must be taken with the same receiver, antenna and preamplifier settings.]

I invited him to ride in my car so that I could demonstrate my equipment, how I could see multiple sources, and locate sources. He was much interested and after some observation he exclaimed, "Why, you have better equipment than the FCC!"

I was pleased at the compliment but knew it was not really true, except that I do have better equipment for specifically locating power-line noise sources. My not-too-elaborate installation is shown in Figs. 4–2 through 4-5.

Driving a short distance, I proceeded to tentatively locate two source structures, making observations on my scope within the vehicle; we then went back to his ham shack where I looked at his equipment and observed the noise on his S-meter, it read about S-9. Fig. 4-17 shows a typical S-meter, indicating a high noise level. The S-readings are shown on the upper scale of the meter.

Fig. 4-17. S-Meter on Ham Receiver.

I advised a power-line crew in the area that I could use some help, and they said they would be over shortly. While waiting for them, I confirmed, out of the vehicle with my hand-held 237 VHF-UHF locator, the two source poles that had been suspected from observations in the vehicle. When the crew arrived, they found that one of the sources was not on the pole at all. It was a short distance out on the neutral conductor of the street-crossing span. Here an automatic splice was sparking. See Fig. 4-18.

This neutral conductor was obviously sagging more in the afternoons because of the increased temperature. This caused looseness in the splice, and it was sparking badly.

To correct the street-crossing noise problem, the line crew replaced the sparking tension-splice with a compression-splice, more suitable for a slack road-crossing span.

The crew then proceeded to locate, with their RE 247 hotstick sniffer, loose and sparking hardware on the pole of the second source. They tightened it up to correct the noise there.

Checking back at the ham shack, we found the noise was down <u>below</u> S-3, and Ernie was pleased.

A.

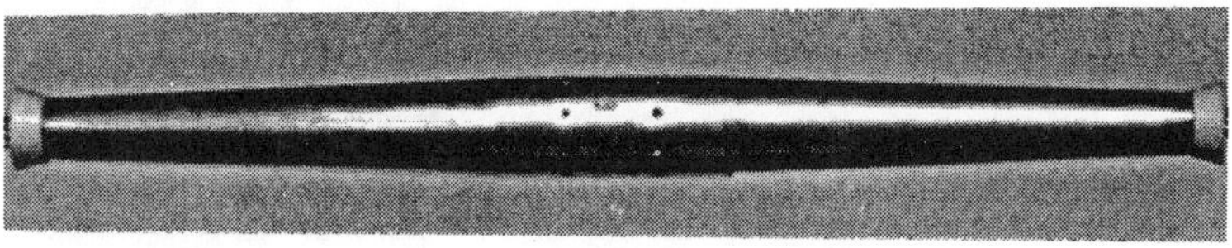

B.

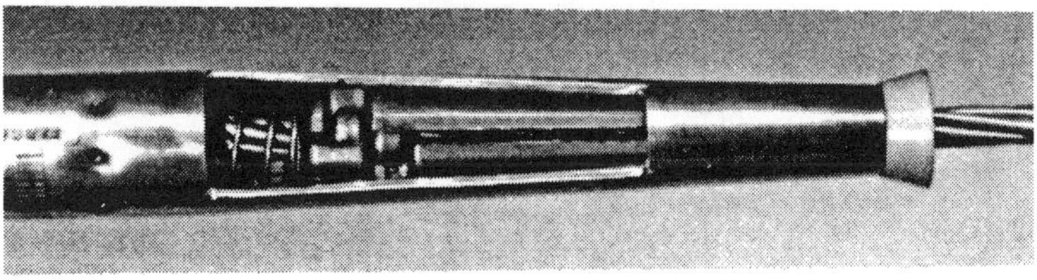

Fig. 4-18. A. Automatic Splice in Conductor Span. B. Cutout View Showing Action of Wedges Which Force a Good Contact of Conductor to Wedges, with Sufficient Tension. Demonstration section **B** from Ron Furrer [6].

Fig. 4-19. Ham Radio Operator's Residence with Several Antennas Including a 10/15/20 Meter "Rotary Beam" Yagi Type [1].

Radio Amateur Frequency Bands.

Radio amateurs often talk of operating on the 10-meter band, the 20-meter band, the 40-meter, and other bands rather than talking of the frequencies they are using. The 40-, 80-, and and the 160-meter bands are especially troublesome from a noise standpoint, In the 80/160-meter bands especially, the operator may hear reradiated noise whose source is at a structure up to a mile or more along the line from the source structure. Never-the-less the author stands by his statement that *most* complaints from radio amateurs result from sources (on or off the line), that are within six blocks of the operator's home.

TABLE 4-2 AMATEUR OPERATING BANDS

Band	Frequency
160 meters	1800 - 2000 kHz
80 "	3500 - 4000 kHz
40 "	7000 - 7300 kHz
30 "	10.10 - 10.15 MHz
20 "	14.00 - 14.35 MHz
17 "	18.060 - 18.168 MHz
15 "	21.000 - 21.450 MHz
12 "	24,890 - 24,990 MHz
10 "	28,890 - 29.7 MHz
6 "	50 - 54 MHz
2 "	144 - 148 MHz
1.25 "	222 - 225 MHz
70 centimeters	420 - 450 MHz
33 centimeters	902 - 928 MHz

Shown in the table are the fairly common "meter" (and "centimeter") bands used by radio amateurs, from 1.8 MHz through 1000 MHz (1 GHz). Many other frequencies are also available for their use beyond this range.

The frequencies shown are those allocated by the International Telecommunications Union for Region 2 (Canada, the U.S., and all of North and South America).

Some frequencies are restricted in use in some geographical areas and may be limited to certain license classes as issued by the member countries.

As you can see, the amateur radio frequencies are well-represented throughout the radio spectrum; however, they actually occupy only a small portion of it. Other services in the spectrum include:

Fixed Service. Between specified fixed locations.

Fixed Satellite Service. Between fixed earth stations and satellites.

Mobile Service. Mobile-to-mobile or to land stations.

Maritime Mobile. Ship-to-ship, or ship-to-land.

Aeronautical. Aircraft-to-aircraft or aircraft-to-land.

Broadcasting. Transmission to the general public such as TV, AM/FM radio, and weather.

In this spectrum you may observe weak TV signals at unexpected locations because of cable-TV leakage. However, this is strictly regulated by the FCC and, if it is over the level of 50 microvolts per meter at 10 ft (3 meters) from the TV cable, it must be immediately corrected. See Chapter 12, *Cable TV Leakage*.

THINGS TO REMEMBER:

(When Locating RFI Sources)

1. RFI includes both radio and television interference (RI +TVI).
2. The RFI source will be at the pole (structure) where you find the highest noise level at the highest frequency.
3. A spark fires and causes noise when voltage breaks down the air across a small gap–the usual cause of power-line RFI.
4. Sparking sources are shorted out by moisture (including morning dew).
5. Loose hardware causes most sparking sources.
6. Sparking results when good contact is lost in the metal-to-metal connections between insulator units because of corrosion and light-mechanical loading.
7. Sparking may also occur between metal insulator caps and porcelain, especially in aluminum-capped units.
8. Close to a sparking source pole, noise is always detected at frequencies from 1 MHz to 500 MHz.
9. TV pictures provide good clues for diagnosing TVI *and RI.*
10. An oscilloscope lets you *visualize* the noise sources and see what's going on "out there."
11. Different sources display different patterns on the scope.
12. You may use television frequencies to locate radio amateur, AM-radio and TV interference sources.
13. To adjust your scope you can use a TV channel pattern for reference.
14. Noise appears as a group of spikes each 60-Hz half-cycle.
15. Pole transformers *are not common* noise sources.
16. You will detect higher and higher noise frequencies as you get closer to a source pole.
17. A source pole may have sites of *radiation* and *reradiation.*

18. Your automobile AM-radio may give you a clue, but it cannot with certainty identify a noise-source pole.

19. Your vehicular oscilloscope and rotatable antenna will usually get you within one or two poles of a source pole.

20. Your UHF locator, operating, for example, at 330, 500, or 800 MHz will identify the source structure.

21. The higher UHF frequencies will narrow down the area of the noise-source observed on the pole.

22. Changing orientation of your hand-held antenna from horizontal to vertical may also help to narrow down this area.

23. The author estimates that the RE Model 250 parabolic locator will, from the ground, pinpoint sparking-sources on distribution lines at least half of the time.

24. However, locating a precise area on the pole is not usually required because the power-line crew will pinpoint the source with their Hotstick Line Sniffer.

25. Power-line noise and noise from thermostats in the home may look very much alike on the oscilloscope.

26. Transmission lines on wood poles are more suspect than those on metal structures.

27. Sparking at the source generates frequencies that are <u>much weaker</u> at the higher frequencies, and these "fade out" rapidly as you move away from the source pole. See below:

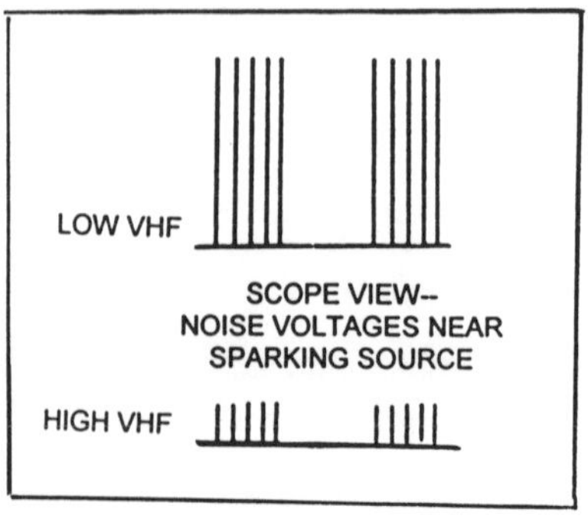

Chapter Five

RFI SOURCES AND CORRECTION

Introduction

Where does *locating* of sources end and *correction* of sources take over? You will find that these are closely related. Correction is sometimes almost incidental to locating. Both are important to the RFI investigator, however, because in addition to locating, he often finds himself directing the *correction activity*.

Years ago, and when power-line investigators were few, the *locating* of sources was considered difficult, and *correction* was considered relatively easy. It was not appreciated that you could really see what was going on out there with an oscilloscope. Also, that if you had a complaint of power-line noise on AM radio at (1 MHz), for example, it was not realized that you would observe this same power-line noise at higher frequencies as you approached the source, and then *be able to locate* it at the structure at 500 MHz (or another UHF frequency). One reason for this is that the usual locating equipment in those days did not operate at UHF frequencies, or even at VHF frequencies.

With present-day equipment and a proper method, *locating* can most often be accomplished easily and rapidly. *Correction* is also often easy. In some cases, though, it may be difficult, for example, when the source location is awkward for the line crew to reach, or if the line must be shut down to do the work.

Examples, RFI Sources. A noisy structure (Fig. 5-1) causing television interference was easy to locate but not so easy to correct:

Fig. 5-1. Complaint-Source Structure. Possible sources here, and their correction are shown in Table 5-1. See Story #20, Chapter 9.

TABLE 5-1. SOME SOURCES AND CORRECTION

See Fig. 5-1

A. Loose or corroded hot-line clamp.
Replace hot-line clamp (clean conductor first).
Use lock-washer-type hot-line clamp.

B. Switch mounting bracket loose.
Tighten hardware.

C. Top unit pin-bracket loose on pole.
Tighten; use lock washers.

D. Insulator tie-wire loose on conductor.
Replace tie-wire after cleaning conductor.

E. Insulator pin-bolt nut and washer loose.
Tighten; use lock washer.

F. Dead-end insulator unit loose.
Replace units, preferably with single-unit polymer type. Or, for temporary correction, apply silicon semi-conductive grease between joints.

G. Crossarm braces loose.
Use longer lag screws or lag studs with lock washers.

H. Cutout switch clamping-strap loose.
Tighten.

J. Device mounting-bracket nut/washer loose.
Install lock washer.

For C, E, G, and J, consider using a double-helix lock washer, as shown installed in Fig. 5-9 page 5-12.

The examples of the specific sources listed are first-hand, that is, these specific sources have been located and/or corrected by the author, or by one of the authors in Chapter 9, *Stories from the Field.* You will see tabulations in the literature listing many causes of interference where some of the sources listed are from third-hand reports–some of these are certainly not common, and are questionable.

A very common loose-hardware condition is that of loose inside-nuts on a double-arming (DA) bolt, as shown between the two crossarms. See A in Fig. 5-2. With drying and shrinking of the wood, these nuts readily loosen.

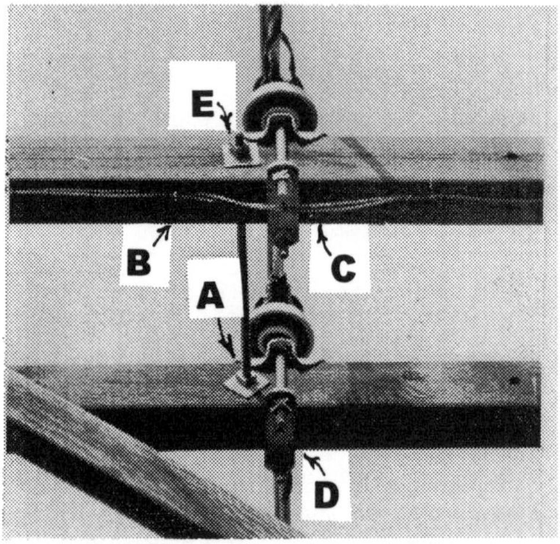

Fig. 5-2. Several Possible Source Locations on Double Crossarm. See Table 5-2.

> By the time a complaint is made to the power company, it is often found by the investigator that several sources are present and are radiating or reradiating from the line into the residence or building of concern

TABLE 5-2. MORE POWER-LINE RFI SOURCES

Shown on Fig. 5-2:

A. Loose inside nut on D-A bolt.

B. Loose metal staples on ground or bonding wire.

C. Ground or bonding wire too close to hardware.

D. Insulator mounting-bracket hardware loose.

E. Flat square washer too close to mounting bracket.

At E replace square washer with round acetyl resin washer, RE #608 ⇒

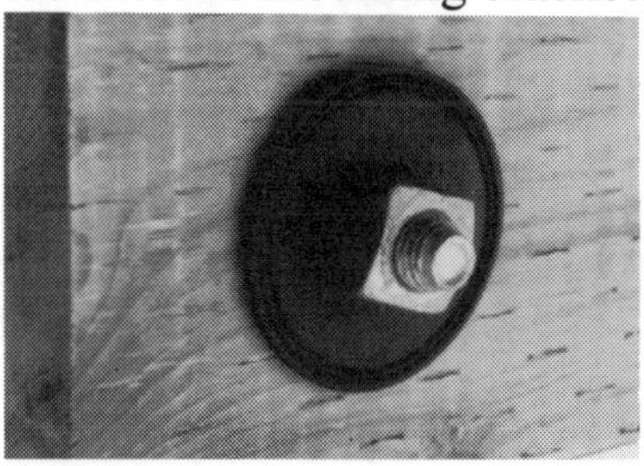

Other Sources:

- Sparking between lightly-loaded vertically mounted porcelain suspension insulators.
- Sparking in lightly-loaded dead-end insulator connections especially in short spans.
- Sparking between insulator cap and porcelain, in aluminum capped insulator.
- Sparking/arcing in crack completely through porcelain in aluminum-capped insulator.
- Side-mount post insulator mounting bracket loose.
- Guy wire touching neutral, or touching ground wire, telephone messenger-wire, or cable TV messenger wire.
- Loose wire or "trash" on conductor.
- Clamp loose in trunnion of "clamp-top" post insulator.
- Sparking sources inside metal-shielded enclosures of transformers, regulators, breakers, and other devices. These metal-enclosed sources cause RI, and, rarely, TVI.

Table 5-2. (Cont.)

- Sparking inside porcelain or polymer/silicon-rubber devices such as lightning arresters.
- Sparking in underground primaries, particularly in load-break disconnects. This RFI may propagate on the underground cables and reradiate from the overhead lines connected to these cables.
- Neutral conductor off insulator, touching metal pin.

Correction Devices and Corrective Methods

> The general purpose of corrective devices is to provide (1) good connections between all metal parts in an electric field that are intended to be connected or (2) good separation between metal parts intended to be separated.

Sometimes we have a choice, and either good contact or good separation will stop the sparking *and* the "noise."

Working over a structure to provide good "electrical-contact" connections or good "separation," as required, will assure that many, but not all, noise sources will be prevented. A line crew should not report that "everything looks good up there" unless they've "looked at it" thoroughly with an RFI detector such as the Radar Engineers Hotstick Line Sniffer.

For sparking in poor connections we can provide an electrical path that will short out the spark gaps: this by tightening hardware; by using proper lock washers; by application of devices such as a spring-shunt between insulator units; by the application of conductive grease in a poor connection; by the hanging of weights on an insulator string to increase contact pressure; or by using bonding braid or straps.

To provide good separation (or insulation between parts intended to be separated) we may move brackets and other parts; use insulating washers or insulating sleeves; use insulating staples; use post insulators or use "single-unit" rather than ball-socket or pin-clevis type multi-unit "strings."

Specific Correction Examples

Sparking in dead-end units. To stop the sparking between the tongue and clevis of units of a dead-end suspension insulator assembly a special wire clip is recommended. This fits around the pin and between the tongue and clevis. See Fig. 5-3. The clip used, and the applicator tool are shown.

Fig. 5-3.
Top: Sparking Dead-End Insulator Assembly.
Bottom: (1) Ohio Brass "Static Clip" and (2) Hot-Stick Applicator Tool, Safety Line 8503Q.

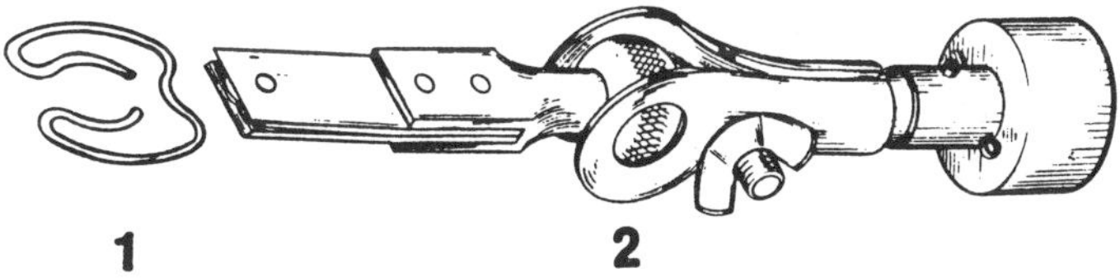

The clip 1 shown is inserted between the tongue and clevis to short out the sparking gap. Sometimes the spacing is large and such clips should then be inserted on both sides of the tongue.

A better practice, of course, would be to change the pair of porcelain units of Fig. 5-3 with a single polymer unit.

Correction with conductive grease. Temporary correction of certain sources can be accomplished by the application of graphite-filled conductive grease. See Fig. 5-4. This is usually effective for two years or more.

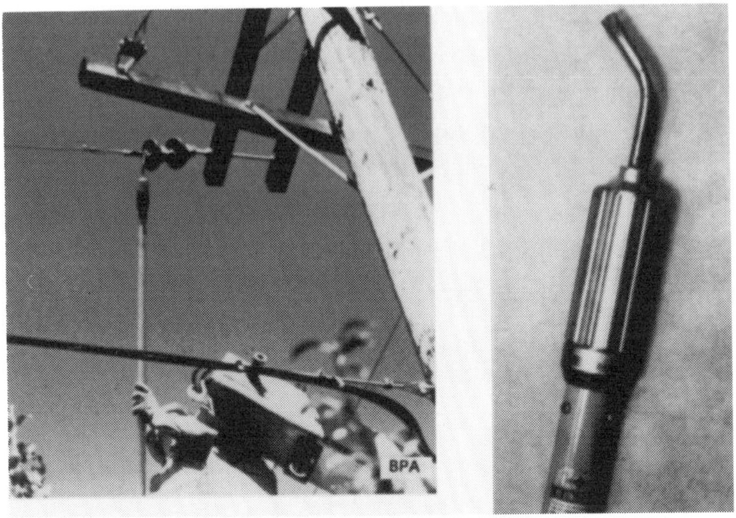

Fig. 5-4. Application of Conductive Grease.
(A). Applying Dow-Corning No. 41 Silicon Conductive Grease, to Short Out Gaps in Deadend Unit Connections.
(B). Hot-stick grease gun, Kearney No. 131614-0.

Turning end of fiberglass rod (protruding from the lower end of a hollow hot stick) forces grease out of nozzle.

Correction by hanging weights. Increasing contact pressure by the hanging of lead weights on suspension strings is sometimes done, shown applied to a ball-and-socket assembly in Fig. 5-5. This is a 100-lb, 7"-diameter lead ball used for this purpose.

A.

B.

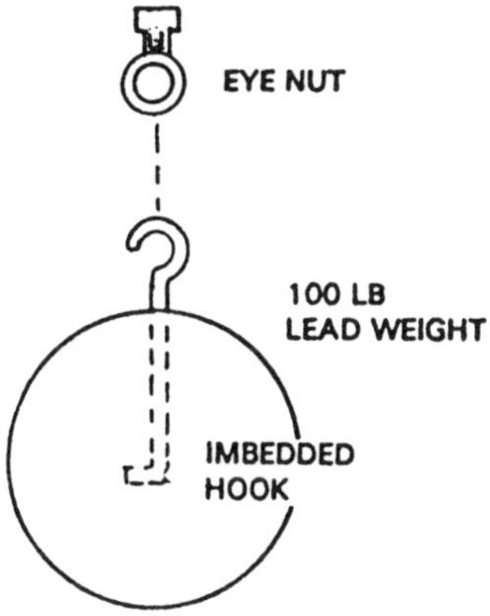

Fig. 5-5. Sparking Elimination by Hanging Weight.

A. 100-lb Lead Weight Hung on Suspension String to Suppress Sparking Noise.

B. Suspension of Lead Weight.

Note that this is a dead-end structure, and that the left section of the line-conductor is connected to the right section by a conductor jumper. This jumper is the only weight on the 115-kV 6-unit insulator string before the weight is hung.

Means of attachment of the lead ball weight to the insulator string is shown in Fig. 5-5B. An eye-nut is attached to a conductor clamp bolt, and the 7" diameter ball is hooked into the eye-nut and suspended by its embedded hook.

Before hanging the lead weight, the insulator units should be rotated one against the other to provide some cleaning of surfaces between the balls and sockets and to improve the electrical contact.

See Chapter 9, Story 20, p. 35, "A Weighty Noise Problem" regarding the hanging of weights.

A unique hardware noise problem. In Figure 5-6, the 230-kV insulator assembly was found to be causing TVI to several homes over a mile from its location as a result of sparking at the conductor-end of the insulator assembly.

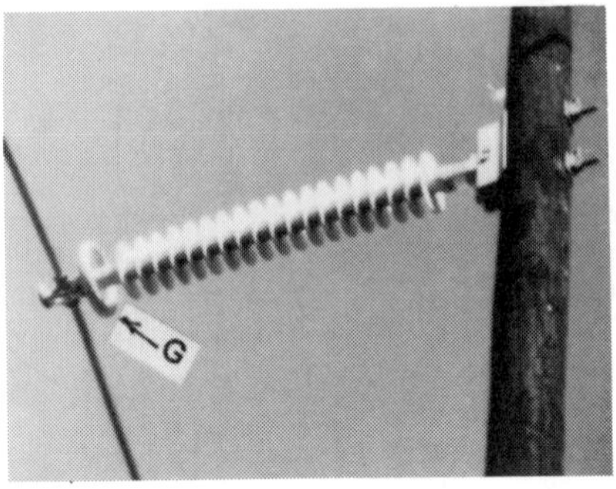

Fig. 5-6. 230-kV Post-Type Epoxy "Rubber-Skirted" Insulating Unit on Wood-Pole Transmission Line.

Here we have a conductor-clamp trunnion on the conductor end. We also have a "grading ring" (G) mounted inboard, and adjacent to, this conductor-clamp trunnion. See Fig. 4, page 1-6 for a trunnion detail.

The purpose of the grading ring is to intentionally distort the electrical field and move its *voltage-gradient lines* outward from, and around, the assembly end. This makes it difficult for partial discharges to reach out for sufficient voltage and cause corona plumes. In this case corona would not form at the 230 kV-line operating voltage.

For practical purposes, these grading rings do eliminate radio noise from *corona*. We would have had a quiet line were it not for this intrusive *sparking*, causing RI *and* TVI from a faulty grading-ring installation. This caused noise radiation and reradiation along at least three miles of transmission line. The Rapid Locating System was put to work, and with its oscilloscope observations, the noise source was confirmed as sparking, and this source was readily located.

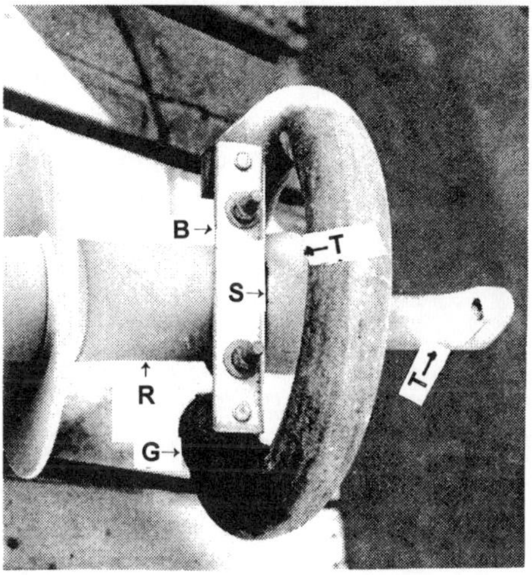

Fig. 5-7A. Mounting of "Grading Ring"G on "Rubber" Sleeve R of 230-kV Suspension-Insulator Assembly.

In the figure, the grading ring G is secured to the insulator rubber sleeve with the galvanized bar B of a U-clamp. Bar B is barely touching the metal-end of the conductor-clamp trunnion T. With the line conductor energized, sparking was taking place at S. Fig.5-7B shows more fully the cup-end of the trunnion which is cemented over the core-end of the insulator unit

If the U-clamp had been located ½ inch inboard from the location shown, this sparking would not have occurred, and we would have had no power-line TVI.

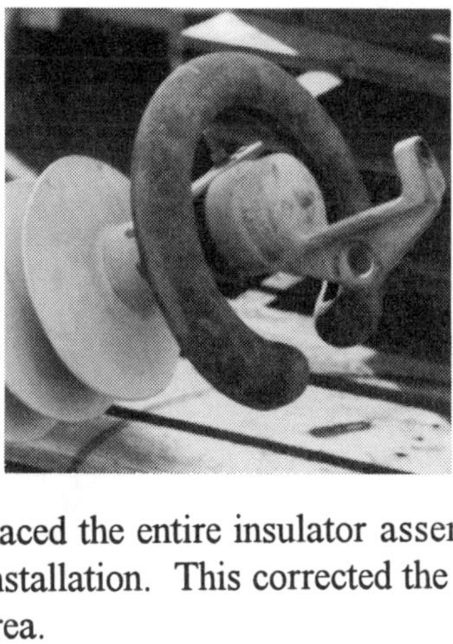

Fig. 5-7B. Attachment of Trunion to Insulator Unit.

The power-line crew replaced the entire insulator assembly with an improved grading ring installation. This corrected the noise and the TVI complaints in the area.

Special lock washers that may be used for hardware mounting on wood-pole structures are the hairpin lock washer, and the double-helix lock washer. These may be used singly or in combination as shown in Fig 5-9. The combination is recommended for noise control on wood poles because of the expansion and shrinking of wood, when wet or dry.

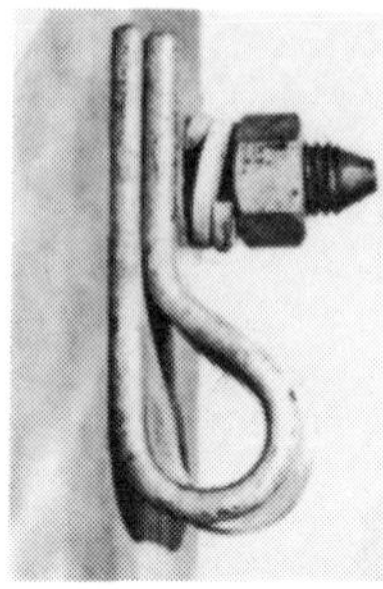

Fig. 5-9.
Hairpin (Leaf Spring) Washer with Double Helix Washer on Wood Pole.

However, The hairpin washer, when loose, will drop down to the position shown and may get too close to other hardware–this lock-washer *is not recommended then* for locations on a pole where there are limited hardware clearances.

On-the-Line "Noise" Devices.

Lightning arresters cause RFI rather often, with internal sparking, and they should be suspected. Fig. 5-10 shows an arrester that was removed from service because it was causing serious TVI. Its noise could be detected at up to 500 MHz. See Story 16 in Chapter 9, and see Chapter 7, page one, about locating noisy lightning arresters.

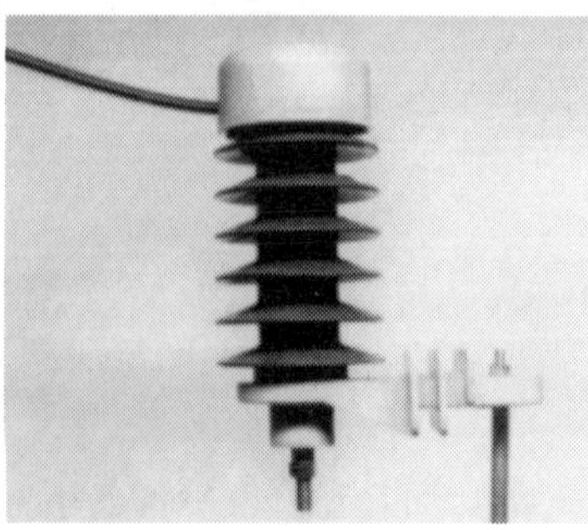

Fig. 5-10.
Noise-Causing Lightning Arrester.

The lightning arrester shown in Fig.5-10 has a polymer housing covered with "rubber" skirts. Porcelain type arresters are shown in Fig. 5-11. In either type, RF noise may occur from internal sparking discharges. Sparking may also occur in loose terminal connections.

Fig. 5-11. Porcelain-type Arresters Protecting 3-Phase Underground Distribution System.

The pole transformer as shown in Fig. 5-12 is another on-the-line device that is often blamed for causing radio and television interference. Yes, it sometimes does (but rarely). A more likely source of noise on the pole shown is the loose nut A and washer holding the bottom transformer bracket. The weight of the device forces, over time, the bottom bracket into the wood and loosens up the mounting through-bolt. Here the lock washer combination of Fig. 5-9 should be used.

Fig. 5-12. The Pole Transformer. Note Loose Mounting Bolt at A.

If a transformer is noisy, it will not generally cause TVI, although loose connections or mounting bolts will. It will cause AM-radio noise. It can be checked at its line-terminal bushings for internal noise, with a 1-MHz hotstick sniffer. See Chapter 7 for description of such a sniffer.

Fig. 5-13.
Typical Fused Cutout Switch.

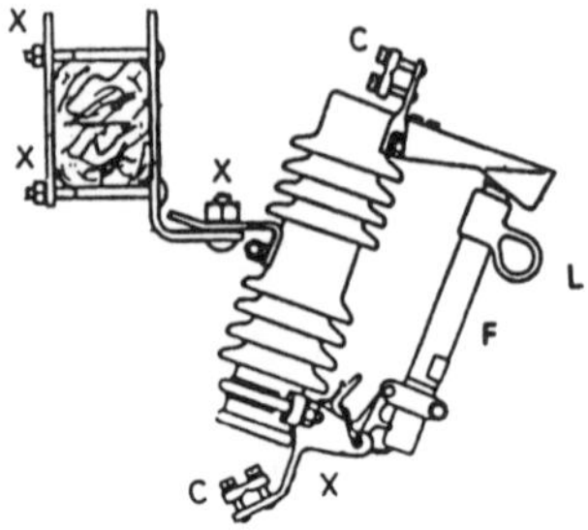

The cutout switch shown in Fig. 5-13 is mounted on a distribution-line crossarm.

Any loose connection such as the nuts and bolts at the X locations shown on the cutout switch can spark to cause RFI. Also, the jumper connector-clamps at the C locations are often aluminum, and if copper jumpers are used, corrosion occurs, with sparking likely.

For a description of the operation of the cutout shown: a loop L at the top of the fuse cartridge F permits the serviceman to open the switch, with a hook-stick, to move the fuse F to a position below the assembly. Further, the switch shown is a dropout type, and the fuse will drop to the open position if the internal fuse link within tube F blows.

> Once an RFI source is located on a *distribution* line, correction can usually be accomplished easily; it can most often be done with the line energized.

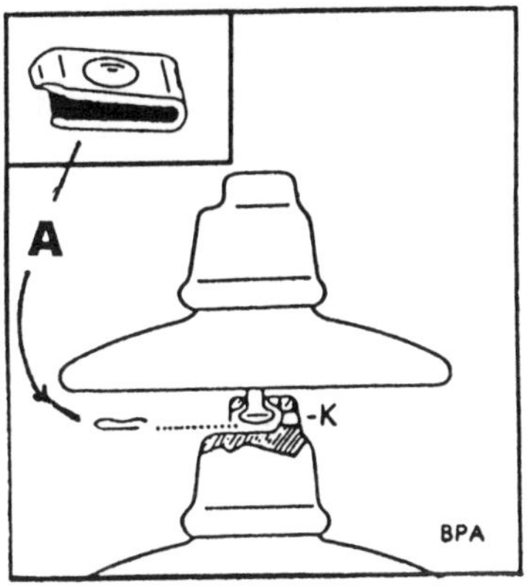

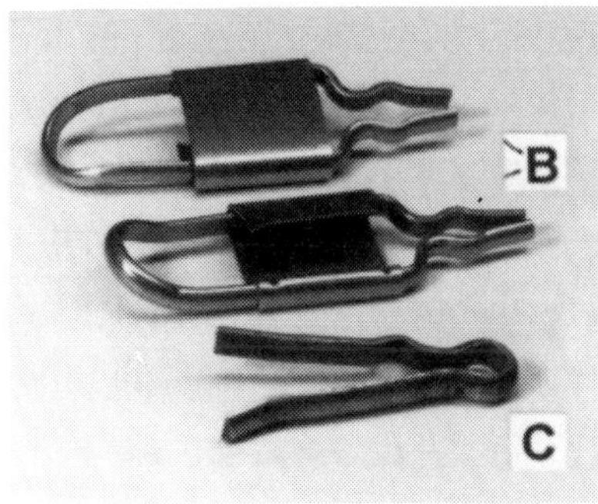

Fig. 5-14. For Prevention of Sparking Between Ball and Socket. Connections, Two Different Devices Available Are: A, the "Corona Clip," and B, the "Hairpin" Cotter Key. The normal installation key is shown at C.

The "corona clip" of A in Fig. 5-14 is shown ready to be inserted after the cotter key has been removed. This safely replaces the cotter key in vertically suspended strings, but one power company reports that the clips *can* break when the insulator strings are mounted horizontally. This clip is a Kearney No. 19815. Calling this device a corona clip is obviously a misnomer, as it corrects sparking and not corona.

Alternately, the "hairpin" cotter key of B (lower) is inserted between the units with its narrow spring-end forced through the cotter key hole K in the socket-cap. This key is a Slater Steel No. 60069. It is found that some insulators have a cotter key hole too small for this clip, and you should check with the Slater Company before ordering.

Stainless steel bonding *brushes* (not shown) have also been used, but are not recommended. These are effective for perhaps a few years, but when the bristles flatten out they cause RFI rather than correct it.

Loosely-hanging porcelain units as were shown in Fig. 5-3 are frequent sources of RFI. The metal connections between units are always at least slightly corroded and are often found to be sparking.

A correction for these is the installation of composite insulator (one piece) units as shown in Fig. 5-16. These usually work well, but there are still connections at both ends of the units which can spark. At the more noise-critical end A, it is suggested that in an installation such as shown the pin be replaced with a threaded bolt, longer than the pin, with its nut thoroughly locked down and its end-threads jammed.

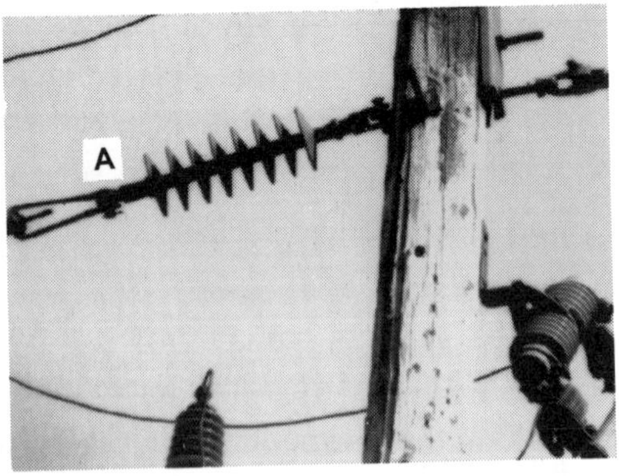

Fig. 5-16. Use of Single-Unit Insulator to Eliminate Sparking.

Another correction for loosely-hanging units is the use of preformed dead-end ties as shown in Fig. 5-17. Here the Preform Line Products "Slack Span Deadend" tie shown is applied. The insulation used in this tie is black aircraft-type neoprene tubing. Linemen like these because they are easy to apply with the line hot. However, a number of failures have been reported in an area of hot, dry summers and cold winters, with sparking and noise resulting.

Fig. 5-17.
Slack Span "Deadend Tie" Installed; Note Neoprene Tubing Separating Tie from Porcelain.

In the crossarm, vertical pin-post insulators are commonly used and sparking may be caused by looseness in pin-to-porcelain or pin-to-thimble connections. Also, sparking may occur between thimble and porcelain. Threads used on insulator pins and in an insulator socket are shown in Fig. 5-18.

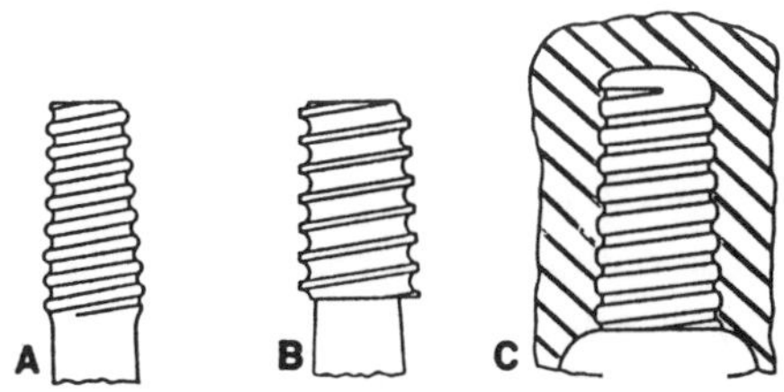

Fig. 5-18.
Threads on Pins and in Porcelain Insulator.

A in Fig. 5-18 shows a plain galvanized threaded steel pin. **B** shows a steel pin with lead-covered threads. **C.** shows the internal threads in a pin insulator unit. If **A** is used, it requires a threaded thimble insert in the porcelain. A good combination is a lead-coated pin B threaded into an insulator unit like C.

Insulators may have factory-installed thimbles and some have semi-conductive glaze on the internal threads.

The hot-line clamp is a common noise source. In Fig. 5-19 a stirrup is used for the clamp mounting. This prevents damage to the conductor. The stirrup is attached to the conductor with compression sleeves. The stirrup *does not* eliminate noise sources which occur at the clamp-to-stirrup connection. The hot-line clamp should be of a type that contains a lock washer, to prevent possible looseness (and noise).

Fig. 5-19. General Purpose Hot-line Clamp on Stirrup.

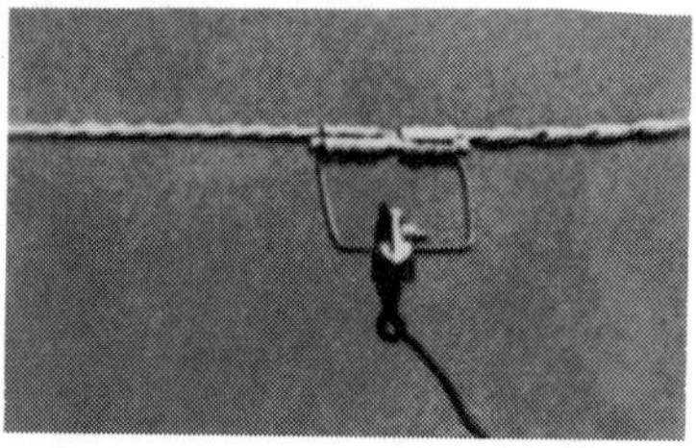

One purpose of the hot-line clamp, of course, is to permit a lineman or serviceman to use the clamp as a switch, to remove the power-line voltage from a device (such as a transformer). He does this with a hot-stick tool which permits him to unscrew the clamp and lower it while the line conductor is energized.

A main purpose of the stirrup is to prevent burning or other damage to the conductor by repeated removal and application of the clamp to the "hot" distribution-line conductor.

Another common source of RFI on wood poles is sparking from loose metal staples to ground or bonding wires. Some good corrections are illustrated in Figs. 5-20 A and B.

Fig. 5-20 A.
Fence-Type Galvanized Staples with Spark-Suppressing Plastic Insert, for #6 or #8 Bare Copper or Aluminum Wire.

LINSDEY-SS-8-15

Fig. 5-20 B.
Square-Type Staple with Insert for #4 Wire.

LINDSEY TVI-SS-4

Off-the-Line Sources

Many power company RFI investigators will agree that the three most common off-the-line RFI sources (which reradiate from the power line) are **(1)** the doorbell transformer, **(2)** the heating pad or blanket, and **(3)** the fish-tank heater. Whereas the dead-end porcelain bell insulators are known as a notorious noise generator for on-the-line sources, the doorbell transformer may be classified in the notorious category for off-the-line sources.

> The dimmer switch and the touch control are also notorious. These are *not sparking* sources. They cause RI but no TVI. Their noise *does get on* the power line and reradiate for a few spans. Such switches generate jagged waveforms which *appear* to be fairly sharp, but their rise-times in no way compare to those of sparking.

(1) The doorbell transformer is a noise generator because of sparking in its thermostat. In Fig. 5-21A the thermostat is shown lifted out of pocket for photography.

Fig. 5-21A. The Doorbell Transformer.

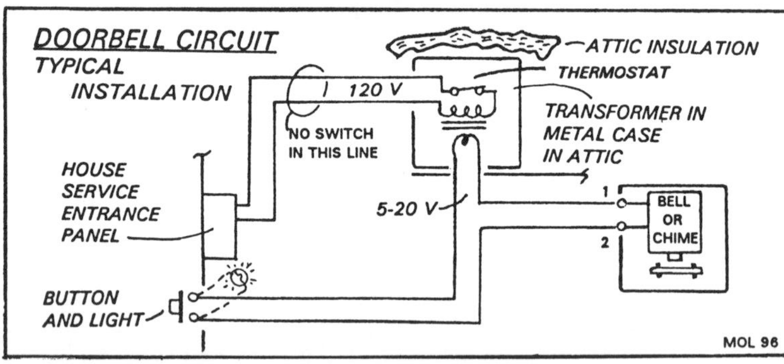

Fig. 5-21B. Typical Circuit of Doorbell Transformer.

The transformer in Fig. 5-21B is shown enclosed in a metal case, and it is often found in an attic covered with insulation. Thus the transformer maintains a fairly steady temperature, and in the summer, often near that required to open the thermostat contacts. The transformer is always heating, to some extent, with the thermostat in its normally closed position.

Note that the 120 V input line comes directly from the house entrance panel, and cannot be shut off except with the entrance circuit breaker to which it is connected...and the RFI generated by a sparking thermostat has a rather direct line, for noise conduction, to the entrance panel and then to the outside world by reradiation from your company's power line.

In investigating a complaint where you suspect a doorbell transformer: if the button light (see drawing) is flickering, you have found the noise source, and the doorbell transformer thermostat is opening intermittently and sparking. Do not press the button right away, as that may stop the thermostat cycling...wait until the homeowner can observe it while monitoring with your 1-MHz sniffer (*See Chapter 7*) or your Model 247 hot-stick sniffer.

(2) Another example of an off-the-line source is in the partially-opened "thermostat-contacts" of a heating pad. The resistance heating-element of such a pad, as shown in Fig. 5-22, is used in the "Heat Wrap™" pad, and this example was very noisy. Here the #1 thermostat bimetal contacts were sparking.

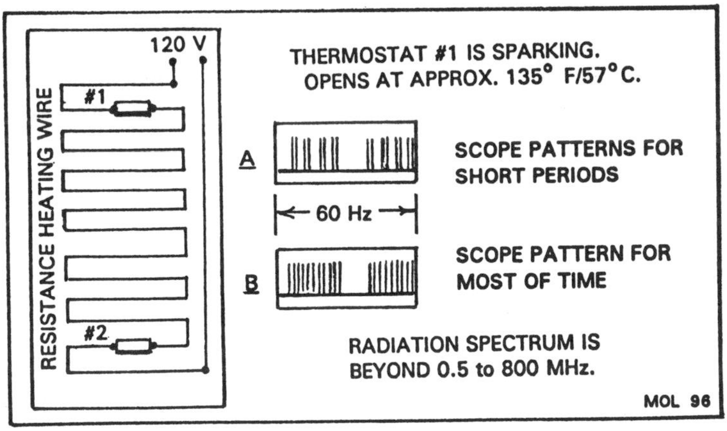

Fig. 5-22. Schematic of Noise-Causing Heating Pad.Items #1 and #2 are enclosed thermostats.

In the figure, the thermostats are shown near the heating-wire ends; these are enclosed in small metal housings. In tests, these thermostats were each found to open at about 135°F/57°C. Scope observations showed a temperamental character of the sparking, as seen in its noise patterns. During some short periods the pattern of (A) was observed, but most of the time the pattern of (B) was seen. In (A) regular "double spikes" are seen in the spike groups. However, such an observation in the field could tell you that you are looking for an off-the-line thermostat-type source. The author has no explanation for the sub-grouping of (A) except that contact bounce is possibly occurring in the thermostat of (A). [RFI radiation here exceeded 0.5 to above 500 MHz.]

> The power-line interference investigator is often led *by the power line,* using his locating equipment, to sparking sources located not only "on-the-line," but also to sources "off-the-line."

(3) A third and common troublesome off-the-line device is the fish tank heater of Figure 5-23. The cause of the noise in this example is sparking in the thermostat contacts.

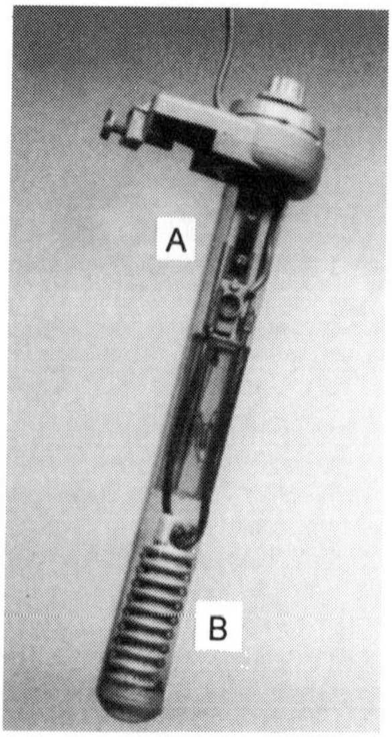

Fig. 5-23.
The Fish-Tank Heater; Its Bi-Metal Thermostat Is Shown Near the Top in the Glass Tube, (A) and the Coiled Nichrome Heating Element (B) is in the Bottom.

A signal-type off-the-line (non-sparking) source that you may observe is shown in Fig. 5-24. This is TVI observed from a nearby microwave oven.

Fig. 5-24. TVI Observed on Channel 4 from Microwave Oven.

Table 5-3 lists further *off-the-line* sources that the power-line interference investigator may encounter in a complainant's home. When such sources are found, it is well if you can advise the customer just what the problem is, what to replace, and where to start in getting it corrected.

> Correction of sources should generally be done only in fair weather when the power-line noise is present. Then the *corrective action* can be monitored by the investigator's equipment at ground level, or by the line crew in a service-truck bucket with the hotstick line-sniffer.

TABLE 5-3. OFF-THE-LINE RFI SOURCES

Sparking Type

A. Non-Power Line Device Sources, Locations,

1. Doorbell transformers.
2. Electric blankets; heating pads
3. Fish tank heaters.
4. Other device thermostats.
5. Ignition interference from vehicles and lawn mowers; vehicle flashers.

1 through 5, above, are more-or-less in order of objectionable occurrence. Others below are listed in random order:

Electric fences.
Yard or street lights (controllers or lamps).
Flashing signs and decorative lights.
Neon light connections or hardware.
Decorative lights.
Furnace controllers.
Furnace igniters.
Air purifiers (cause RI only).
Fluorescent lights. These *usually* cause RI only.

Non-Sparking Type

B. Non-Power Line, Signal-Type RFI.

Dimmer and touch switches. These usually cause RI only.

Radio or TV interference from vehicular, aircraft, or amateur or CB radio.

Cable TV leakage.

Defective TV antenna preamplifier oscillation or other TV amplifier radiating TVI.

TV sets on same TV cable without proper splitter.

Overloading of TV set from nearby TV or FM stations.

Nursery monitors.

TVI from a VCR, tuned to a particular channel.

TVI from a microwave oven.

RFI from personal computers.

SUMMARY

- The locating and correction of sources are closely related.
- Correction of RFI sources on a power line is often easy, but may be difficult and time-consuming if it is found necessary to shut down a line for correction..
- The power-line RFI investigator often finds himself directing the correction activity.
- Correction of sources should generally be done only in fair weather when the power-line noise is present.
- Pole transformers are seldom noisy, but when found to be, usually cause RI and no TVI.
- Loosely hanging dead-end porcelain bells are notorious noise generators.
- Pin insulators may have internal sources such as metal-thimble-to-metal-pin sparking.
- Loose hot-line clamps are common sources.
- Very common *off-the-line* sources are the doorbell transformer, heating pad, electric blanket, and fish tank heater.
- Dimmer switches and touch switches are the non-sparking type and cause RI *but not* TVI. The radio interference resulting *may* propagate on the distribution line for a few spans.

Chapter Six

NOT-SO-RAPID LOCATING for RADIO AMATEURS & HOME-OWNERS. FCC CONCERNS

Introduction

In this chapter the author suggests what radio amateurs or other home owners might do, with minimum equipment, about their interference problems before calling the power company for help.

In some cases you will be able to find and correct the noise causing the problem after making a few simple tests with simple equipment, as will be described. Most often you will be able to tell whether or not the noise is coming in via the power line.

In any event, when you call the power company, if that action is indicated, you will be able to provide them with helpful information.

We start the discussion with a simple method using only an AM/FM "pocket" transistor radio and making observations on that, and with making observation on your living room TV.

We will also discuss two other not-so-rapid methods requiring some additional equipment. Then, if you find the noise is coming into your home via the power line, you should be able to find the *general location* of the noise on the power line–that is, to localize it to within two or three poles.

The Transistor Radio Method (1)

A method that may work for you, particularly if there is only one source out there, is here described. This method is in cartoon form and is from a booklet prepared by Southern California Edison, and is used with their permission.

Here it is assumed that the power lines are overhead distribution and that the complainant lives near such a line.

Fig. 6-1. *Having a hard time receiving your favorite AM radio stations?*

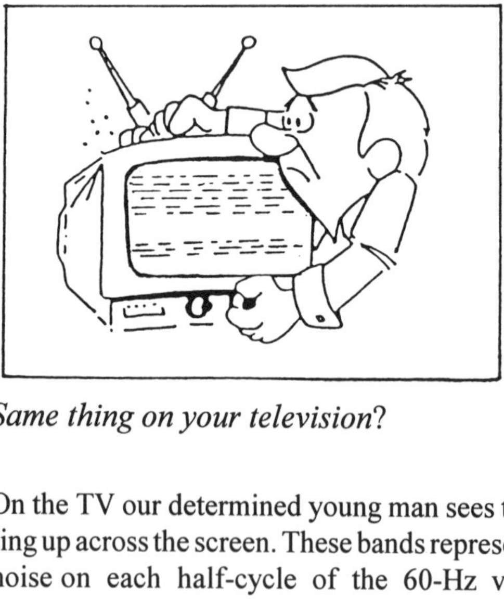

Fig. 6-2. *Same thing on your television*?

[**NOTE**: On the TV our determined young man sees two "bands of noise" drifting up across the screen. These bands represent "electrical sparking" noise on each half-cycle of the 60-Hz voltage. This electrical noise *may* be power-line noise *or* device-noise in the residence.

If the noise-bands are seen on low-band TV channels 2-6 only, the noise is coming in from a distance and probably from the power line.

If the bands are *also* seen on the high-band channels 7-13, the source may be from very close power-line poles, but most likely it is from a device source in the residence. It is assumed here that the TV has its own antenna and is not on cable...ed.]

In Fig. 6-2 it is seen on the TV that the two bands of noise are unequal in width. This is often the case, and sometimes the second band is even completely missing. This is because of peculiarities in the noise-making spark gap, where sparking-spikes fire more readily in the positive-going half-cycle than the negative (see Chapter1, Fig. 9).

Fig. 6-3. *You have what's called RFI (Radio Frequency Interference–RI & TVI) and it may be coming from your home!!!*

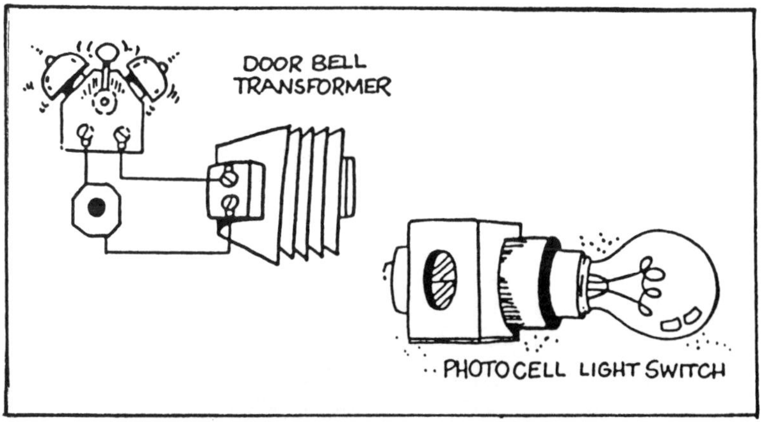

Fig. 6-4. *You have potential interference makers all over the house. The door bell transformer, the photocell light switch and other electrical devices.*

Fig. 6-5. *Or you may have a tropical fish-tank heater or pump, a heating pad, or a faulty door-bell transformer or other noisy devices.*

Fig. 6-6. *You also could be bothering your neighbors' radios and televisions.*

__Fig. 6-7.__ There is a simple test that can be made. Go to your main circuit breaker box. Listen to the interference on AM--on a pocket transistor radio. Turn the main breaker off. If noise goes away, the problem is coming from your home!

__Fig. 6-8.__ If there is still interference, the problem may be coming from a neighbor's home. But if the problem is in your home: ***1.*** *Turn the main circuit breaker on;* ***2.*** *If it is intermittent, wait for the interference to start again;* ***3.*** *Then do the same test with sub-breakers, one at a time. This will identify the circuit for the device causing the interference. Turn all breakers to "on" except the one that stops the noise.*

Fig. 6-9. *Now find out what is on that circuit. Check fish tanks, door bells, heating pads, electric blankets, dimmer light switches, photocell light switches, sonic pest control devices, and any other electric-powered devices that may be plugged in on that circuit. When you find the noise–maker, leave it unplugged and advise the homeowner what he might do to correct it.*

Fig. 6-10. *You've just become the hero of the neighborhood!! No more interference on the radio or television in your home or the neighbors'.*

[Note: In actual practice it is not necessary to extend the whip antenna as shown in the cartoons because this whip is for FM reception only...ed.]

RFI-Inspector Method (2)

This method uses improved equipment.

At a power-line noise workshop presented by the author and Mike Martin of RFI Services, an FCC inspector in attendance showed the author a locator that "they used" for finding power-line noise. This was a transistor radio using a microammeter connected to the radio output. A meter reading provides a much better indication of noise level than sound output, as you walk along the power line comparing readings from pole-to-pole.

Here, of course, you have a rudimentary field strength meter --one with very limited frequency range, and uncalibrated, but capable of taking readings for comparison purposes. Such a modified radio, as shown in Fig 6-11, may be obtained from mike@rfiservices.com

Fig. 6-11. The RFI Inspector.

In using this "FSM," hold it a few inches from the pole ground wire (which runs up the pole and is stapled to it). Tune to a station-free location on the dial and adjust the output for a fairly low level of the noise heard and seen on the panel meter. Do not change the control settings as you move from one pole to another until after you have made the comparison. If the reading goes up, reduce level and proceed to the next pole.

[NOTE: All poles may not have ground wires, but when present these provide good reference points for comparing noise levels.]

An interesting observation that you may make is that a normally very weak radio station will often come in much stronger if you hold this radio near a distribution-pole ground wire. Here the power line and its ground wire are acting as a receiving antenna.

From-the-Vehicle Method (3)

With a ham radio receiver in your vehicle, observing its output meter (S-meter, Fig. 4-20) and monitoring the noise using the AM detection mode, you can close in on the noise source by going higher and higher in frequency. Here you may use a Radio Shack #21-012 vertical whip antenna with a magnetic-base mount, on the roof of your vehicle. This antenna has a receiving frequency range of 30 to 300 MHz and is very suitable for this method. This antenna is shown below.

Fig. 6-12. Antenna used for "Not-So-Rapid Locating."

With this method you would start at your residence, assuming that you are the "complainant," and listening to your receiver tuned to about 30 MHz proceed along the power line, in one direction or the other, until you can observe the noise at

50 MHz. You may, of course, be able to detect it at 50 MHz (6 meters) on the power line when you first leave the residence.

Your immediate goal is to get to the point where you can observe the same noise at 150 MHz (2 meters). You may be going the wrong way, or have made a wrong turn and need to turn around in order to keep increasing the frequency for noise reception.

When you can observe the noise at 150 MHz you still may be a pole or two off, so you should make a couple more passes, and carefully compare meter readings. If you continue to get the same readings for two or three poles, you have found the area.

If there is a transformer on one of the poles located, then check out the residences connected.* You may find a fish-tank heater (or other thermostat device) in one of these houses to be the whole cause of the problem. If these houses check out OK, then it is is probably a power line source.

When you do identify a noisy two-or-three-pole area, it would be wise if you would cruise around to see if you have overlooked another source. If you should get another noise peak at 150 MHz in another area, say five poles away, then you have two sources.

This method, if several sources are present, may be time-consuming, but with patience will permit you to find noisy poles on a distribution line within a pole or two. This is sufficiently close for you to call the power company. To make it real easy, you would need an oscilloscope in your vehicle.

*During the above procedure, if one of the residents is uncooperative, and you are not able to make tests in his home to find a source, then let the power company handle it.

The Power Company's Responsibility

Federal Communications Commission (FCC) rules *do apply* to interference from overhead power lines, and they do expect power company help, if their lines are at fault.

Prior to the investigation of one complaint, the power company concerned received the following letter (in part) from the FCC:

To Manager, Municipal Electric Company,

Dear Mr. Rogers,

The FCC is charged with regulating the radio spectrum to make it as interference-free as possible. Power lines are regulated under Section 15.25 of the Commission's Rules as a device that unintentionally radiates radio frequency energy during the course of its operation. The Commission allows the operation of such devices as long as the emitted radio frequency energy does not cause harmful interference to any licensed radio communication service. Should harmful interference occur, the operator of the interference device must promptly take steps to eliminate the harmful interference.

Please contact Mr. Howard Jones and work with him in resolving this problem. A brief report summarizing the result of your efforts should be submitted to this office within thirty days of receipt of this letter. Your prompt cooperation in this matter would be appreciated.

Very truly yours, *(Signed by the FCC District Engineer in Charge.)*

The FCC, in the case of a particular complaint, may weigh a number of factors when considering the extent to which a utility should go in resolving the complaint. Some of these factors are:

1. The number of persons being interfered with,
2. The number of stations, television channels, or frequencies being interfered with,

3. The severity of the interference; that is, whether it is an occasional annoyance, or whether all channels and frequencies are seriously affected much of the time,
4. The distance from the transmitter station to the complainant's receiver, and
5. The cost to the power company to remedy the complaint.

The FCC is found to be reasonable about the effort and expense expected from the power company in eliminating <u>all</u> interference at a certain location, and it is aware that especially in fringe reception areas it may be very difficult to do so.

However, for a power company to report that "our line crews have thoroughly checked over the lines in the area and have found no problems" should not be considered a reasonable effort at resolving a problem. The line should be checked out *when the interference is on* and with equipment that can detect the presence of power line RFI before such a report is made.

FCC Rule Enforcement

The FCC does have an Enforcement Bureau. This is to protect the many communication services including Amateur Radio.

As an example, the penalty received for a serious infraction, resulting from the complaint of an Amateur Radio Club is of interest, and it is detailed in the following news clipping.

[This clipping is from the October 2002 issue of *Conformity,* a guide to regulatory compliance, 948/846-0888, and is used with their permission.]

Unlawful RF Signals Leads to Prison Sentence for Florida Man

An investigation by the Enforcement Bureau of the Federal Communications Commission (FCC) has led to the arrest and conviction of a so called "broadcast pirate" who was illegally interfering with amateur radio communications.

William Flippo of Jupiter, Florida, was convicted on eight counts brought by the FCC, including unlicensed operation and causing intentional interference to amateur radio communications, and has been sentenced to 15 months imprisonment and a $25,000 fine. The conviction and sentencing came as a result of a June trial, following which Flippo was remanded into custody by federal officials.

The FCC's investigation into Flippo's illegal broadcasting stemmed from complaints received from the Jupiter-Tequesta Florida Repeater Group, a 70 member amateur radio club, that Flippo was intentionally interfering with their communications on two different radio bands. The Commission's investigation began in November 1998, and culminated in the trial this summer.

The operation of an unlicensed broadcast station is a violation of the Communications Act of 1934, and illegal operators may be subject to monetary penalties and criminal sanctions, including fines and imprisonment.

In using one of the not-so-rapid methods, when you think you are within one or two poles of a source, a thorough visual check of these poles with the help of binoculars might reveal a probable source at one of the poles. This may be the cause of the complaint. Be sure to check between poles for trash on the conductors that might be sparking.

If the source you are looking for is on the power line, it will very likely be one of the items listed in Table 6-2.

TABLE 6-2. Several of the Most Common Noise Sources Found on Distribution Lines.

- Pole-guy touching neutral conductor or other wire.
- Foreign object on conductor.
- Loose pin insulator nut-and-washer.
- Loose nut on inside or outside of double-arming (DA) bolt.
- Ground-or bonding-wire too close to hardware.
- Transformer bottom-mounting bolt loose.
- Other device mounting bolts loose.
- Loosely hanging insulators in jumper or slack spans.
- Crossarm braces loose at center lag-screw on pole.
- Loose bottom bolt on side-mount post insulator.
- Defective lightning arrester.

Chapter Seven

LOCATING HINTS & EQUIPMENT PROJECTS

Introduction

In this chapter some special locating and correction information, design details, setup information, and construction projects for the radio amateur or other handyman are described.

The lightning arrester problem. Various people have tried for years to find a good method and develop equipment for locating defective lightning arresters, but to no avail. It apparently did not occur to them (1) to look at a lightning arrester with an oscilloscope, and (2) to observe this at 200 or 300 MHz.

A defective lightning arrester is shown in Fig. 7-1, along with its oscilloscope noise pattern. From outward appearances it doesn't look defective–but it is a serious noise generator because of internal sparking.

The associated noise pattern was first observed by Dave Ohlin of Utah Power (Pacificorp), while working on a serious radio amateur interference complaint. After isolating various devices and hardware and observing with the scope, they found the defective arrester and found that it had the pattern shown in Fig. 7-1B. Replacing the arrester cleared up the ham radio complaint.

Sometime later I was working with RFI Investigator Ron McCracken of Idaho Power on a troublesome TVI complaint. I remembered Dave Ohlin's pattern, and based on this and observing from my vehicle at VHF(200) MHz we located three arresters with substantially the same pattern. Replacing all three arresters was required to eliminate the TVI.

Ron reported that he later found (and replaced) numerous noisy arresters using an observation frequency of 300 MHz.

So here we have two very good indicators for locating defective lightning arresters: (1) A scope pattern observation and (2) an observation at 300 MHz.

The indicators apply to the polymer type of Fig.7-1, arresters rated at 8.4 kV or 18 kV. These are very common distribution arresters.

The cause of arrester failure is believed to be moisture getting into the unit. This can happen if the hot-end lug nut is over-tightened and the seal is broken.

Manufacturer's specs for torquing this nut should be followed. Also, if a 3/8" compression-fitted lug is installed on the hot end of the jumper cable, this torque is not as likely to be exceeded–as it might be when squeezing down the aluminum or copper wire of the jumper.

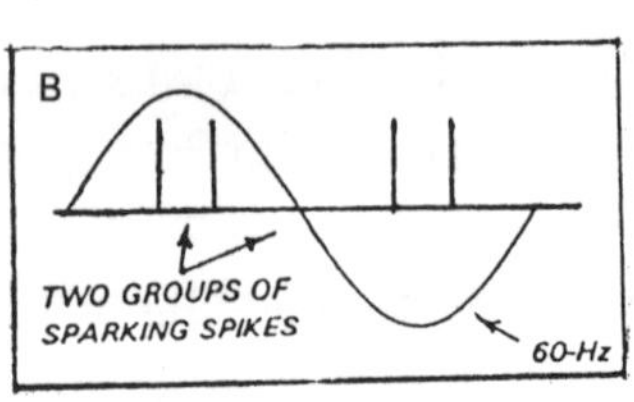

Fig. 7-1. (A) Noisy Lightning Arrester. (B) Its Noise Pattern seen on Oscilloscope.

Some variation in the patterns are seen, but each group, one for each half-cycle, will have *two defining spikes* with a *wide separation.* Fig. 7-2 shows two possible variations. [The 60-Hz reference voltage wave is not seen in the field].

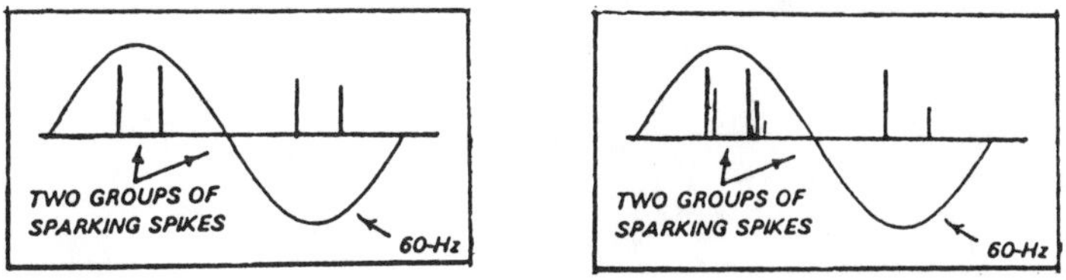

Fig. 7-2. Example Variations, Arrester Noise Patterns.

The notorious distribution insulator unit. Two styles of insulator units frequently seen in the field, often for dead-ending a span, such as a road crossing, are seen in Fig. 7-3. The notorious one is seen in (A). Note that the metal cap of this unit has a more or less square profile, while the better unit (B) has a rounded cap. The cap of (A) is aluminum, whereas the cap of (B) is galvanized iron.

The aluminum cap is apparently not as compatible with porcelain as is iron, and with temperature changes, tends to crack and/or pull away from the porcelain, resulting in very noisy sparks from metal-to-porcelain. So be wary of these frequently used units if seen on your visual inspections.

A

B

Fig. 7-3. Two Examples of Porcelain Insulators.

Fig. 7-4. The 1-MHz Sniffer.

The 1-MHz sniffer is mentioned a number of times in this handbook, nd it is a useful accessory. It is shown on a hotstick on page 4-15. It is a device that is inexpensive to construct.

The sniffer circuit board is mounted in a shielded metal case as shown in Fig. 7-4. Because of this good shielding, practically the only RF detected is by the extended portion of its antenna rod, (Radio Shack 270-1407).

Before installation of this antenna rod, the circuit board from AM transistorr adio RS 12-203 (see Appendix for availability, or alternate), is mounted on the bottom panel of the case (Hammond 1590C, 4.7 x 3.7 x 2.1") using two 2056 x 1" screws, after drilling holes in appropriate places in the circuit board and matching holes in the bottom panel. Use 1/4" plastic mounting-spacers to hold the board off the bottom. The circuit board should hug the right side panel, after making sure the volume control is turned fully on. The speaker is mounted on the front panel, with a grill drilled out similar to that seen on the SC-3 Controller, page 7-6.

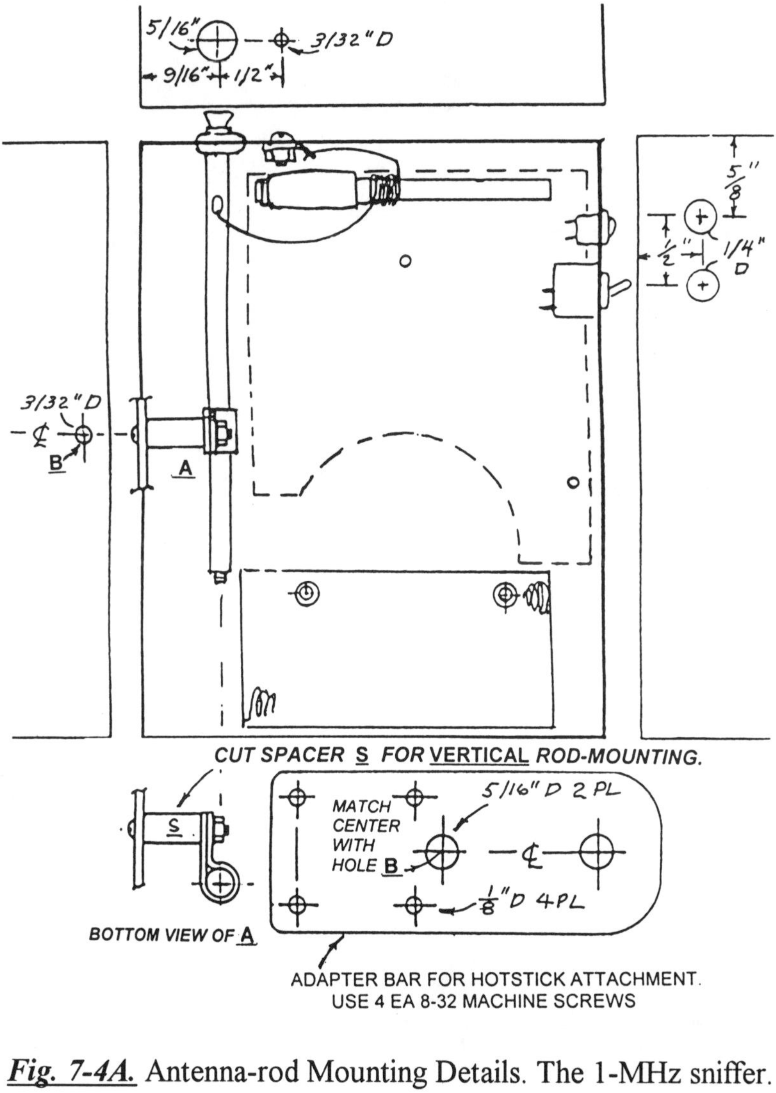

Fig. 7-4A. Antenna-rod Mounting Details. The 1-MHz sniffer.

__Fig. 7-5.__ Photo, SC-3 Scope Controller. Block diagram below.

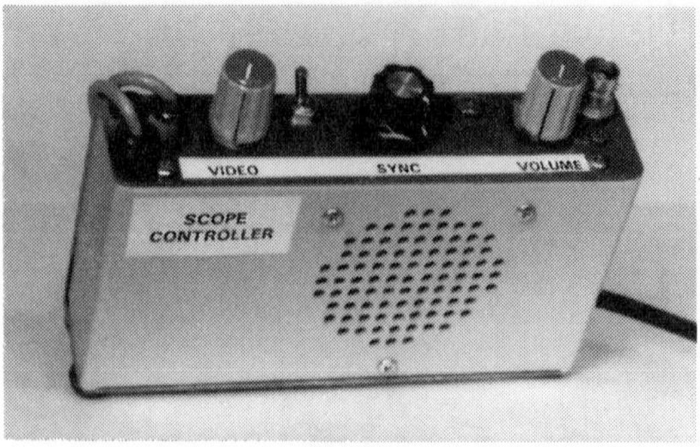

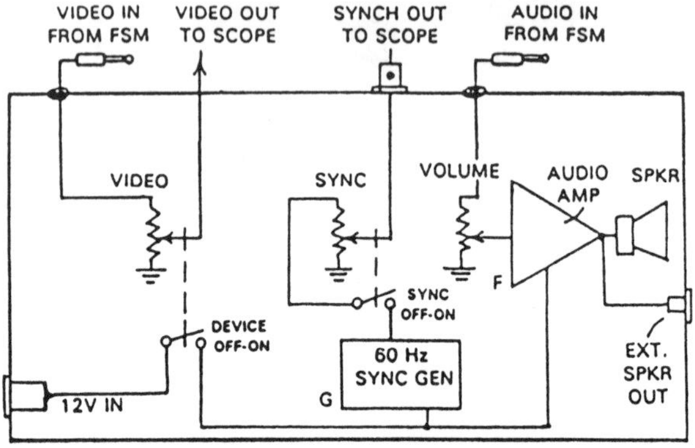

__Fig 7-6.__ Block Diagram of Scope Controller.

The chassis box in which the parts are mounted is a Crown Royal #531 box, 5-1/2" x 3" x 1-1/4." The front panel is drilled with 1/8" holes for the speaker grill pattern. The parts layout is shown in Fig. 7-7.

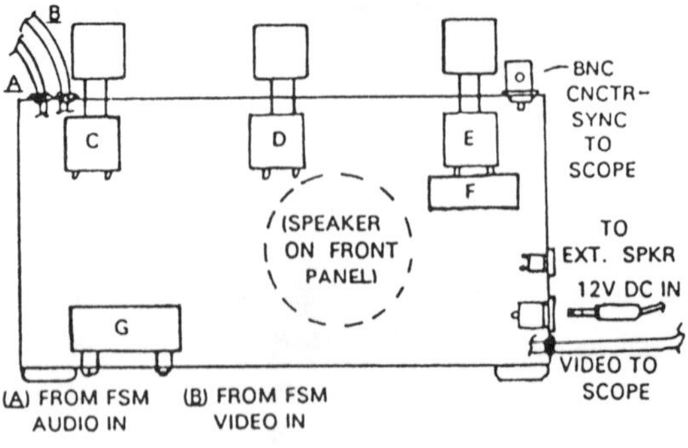

__Fig 7-7.__ Parts Layout for Controller.

Controls C, D and E are 10-k volume controls, RS #272-215. Part F is an audio amplifier (using an LM 386 IC) mounted on the rear of control E on a small circuit board. Part G is the 60-Hz sync (trigger) generator. This is seen in the overlay on Fig. 7-8 and is laid out on a 1" x 2" perf-board and mounted in the box on the bottom panel of photo, Fig. 7-5.

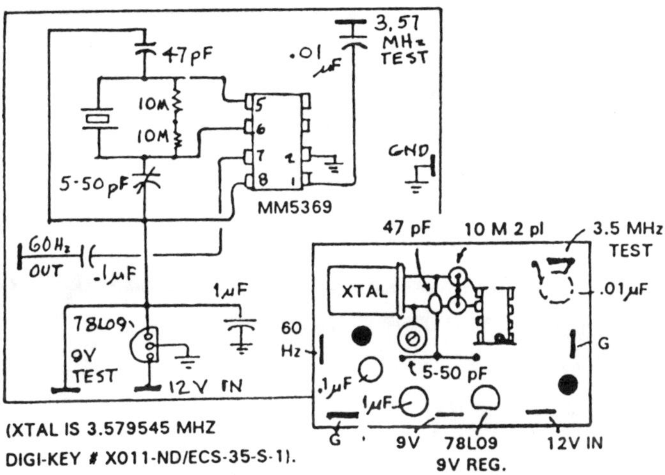

Fig. 7-8. Schematic and Layout of Sync Generator.

NOTE: The MM 5369 frequency divider IC is getting hard to obtain and if you are having trouble finding one, see the Appendix for availability, or for an alternate circuit.

The SC-3 Scope Controller may be used with CRT (cathode-ray type) oscilloscopes, such as the Leader and Hitachi scopes illustrated in the Appendix.

[Note that the toggle switch shown in the photo of Fig. 7-5, is not shown in the block diagram and is not required].

Radar Engineers has available a vehicle antenna-mount with which the antenna can be rotated by hand from an open window. This provides an easy and surprisingly smooth control of antenna horizontal orientation. The indexed slot on the handle provides indication by sight or feel of the antenna orientation. The antenna mount Model 217 is shown in Fig. 7-9.

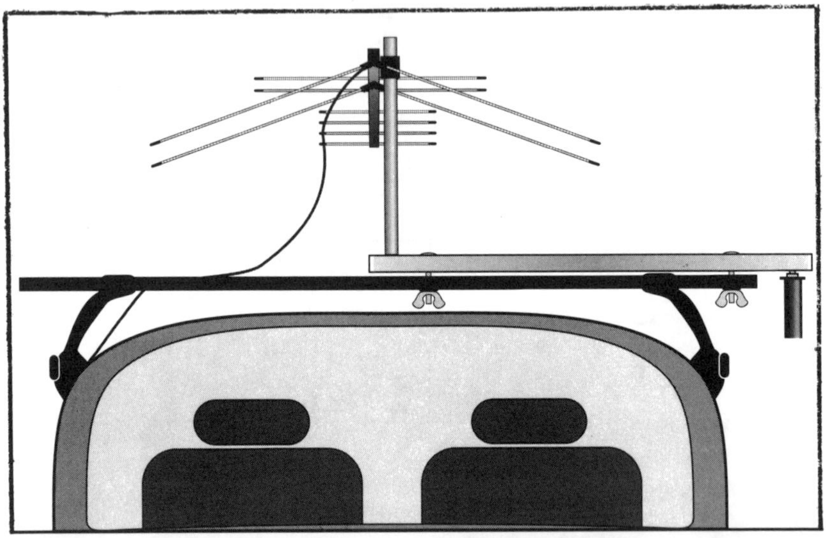

Fig. 7-9. RE Model 217 Antenna Roof-Mount Assembly.

> To rapidly identify a noisy power-line pole (or structure) you need a rotatable antenna on your vehicle, an oscilloscope and a UHF locator.

A half-wave dipole antenna for making standard measurements of VHF signal, noise and cable-TV leakage can easily be constructed, as shown in Fig. 7-10.

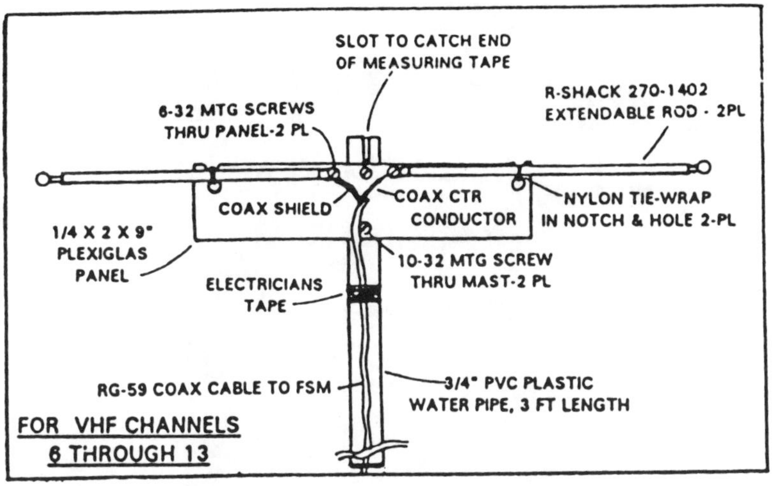

Fig. 7-10. Details, Half-Wave Dipole Antenna.

> If you are a power-company investigator, and you can tell the TV-cable company that on pole number 20-561, for example, you find a cable-TV leakage of over 50 microvolts [as per your measurement as described on page 12-10], you will definitely get their attention.

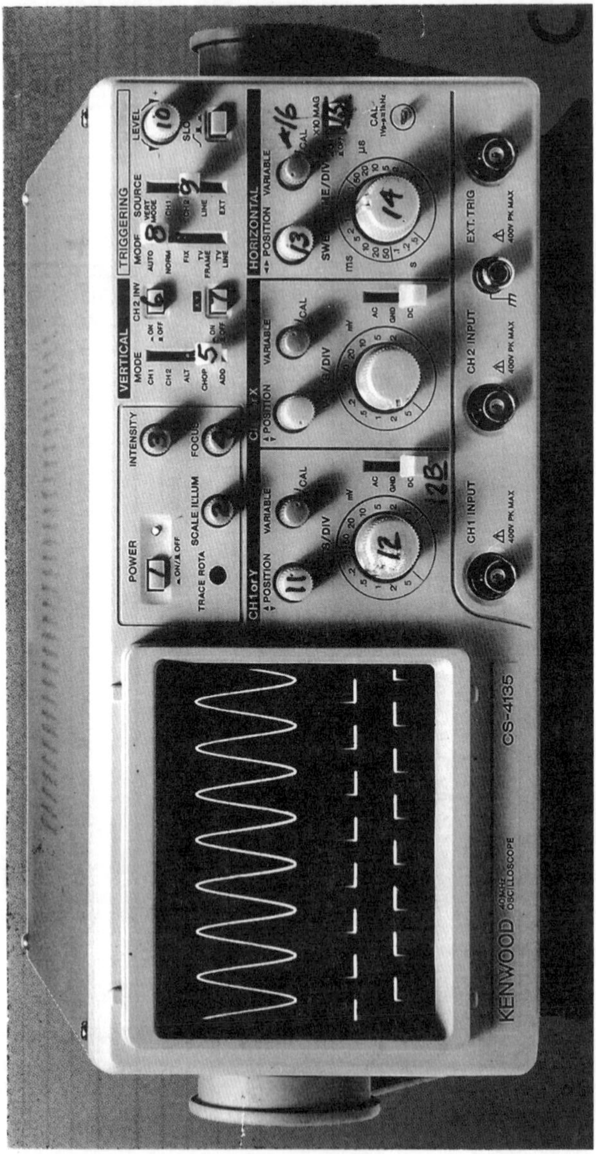

Fig. 7-11. Settings for Noise Locating, Kenwood CS 4135.

Table for Fig. 7-11. Locating Settings for CS 4135 Scope.

1. Push in-out and wait up to 20 seconds for "ON."
2. Start in down position for "dim."
3. Start at 2 o'clock position, then adjust.
4. " " " " "
5. Move toggle to top position (CH 1).
6. Button should be out.
7. " " " "
8. Toggle to top position (AUTO) to start.
9. Move to "LINE" position, or to "EXT" for external use.
10. Start at 12 o'clock; adjust for steady triggering.
11. Start at 12 o'clock position.
12. Start at 1 volt per division.

12B Move toggle up to "AC."

13. Adjust to center trace horizontally.
14. Set at 1.0 millisecond and adjust variable control (16).
    For two noise groups (a sweep of 60 Hz).
15. Button out.
16. (See line 14).

The field strength meter FS-4VU used by the author with the Hitachi V509 oscilloscope is shown in Fig. 7-12. The interface shown between the FSM and the scope is the RE 227, but the author is now using the SC-3 Controller of page 7-6 for this.

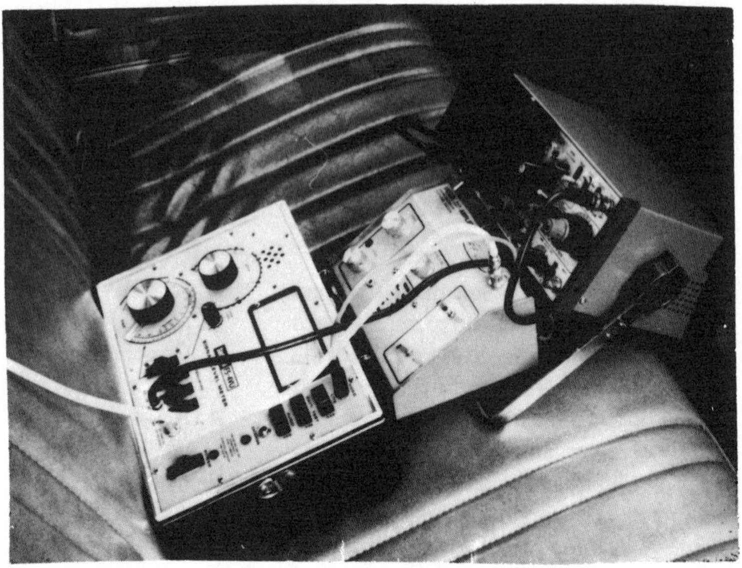

Fig. 7-12. The Sadelco FS-4VU (left) as Used in the Rapid Locating System.

Sadelco, Inc. has manufactured many models of field strength meters (FSMs). Some of these they identified as signal level meters (SLMs). Some of these do not have video output jacks as needed for connection to an oscilloscope. However, most with the prefixes FS-3, FS-4, and FS-5 *do* have video outputs and are suitable for use in the Rapid Locating System. These work well with the SC-3 controller.

Many of these Sadelco FSMs have been used by cable-TV companies and by power companies, and some should be available on the used market. If you are looking for a Sadelco FSM be sure it has these two features: (1) It operates at both VHF and UHF frequencies, and (2) it has a video output.

An excellent antenna for vehicular noise locating is shown on the roof of the author's car. This is shown in its travel position, particularly for non-freeway driving. It is shown in its completely retracted position and is seen to hug the roof. With proper precautions it can be safely used in power substations. This is a modified Channel Master 3619 antenna and has a high front-to-side ratio, and thus exceptional directivity in the 100 to 200 MHz range. Please see the Appendix for information on this antenna.

Fig 7-13. The Vehicular Antenna.

Under lab tests by the author, the voltage applied across this unit was increased beyond 8-kV, and observations in the dark revealed a dramatic ring of brush-like sparks between the bottom of the metal cap and the top porcelain skirt of the unit. As the voltage was slowly increased to 12-kV, a sudden and explosive sound occurred and I, of course, immediately turned off the voltage and turned on the lights. The porcelain was found fully cracked to the core and from top-to-bottom as shown!

Fig. 7-14. The Notorious Aluminum-Capped Insulator That Failed.

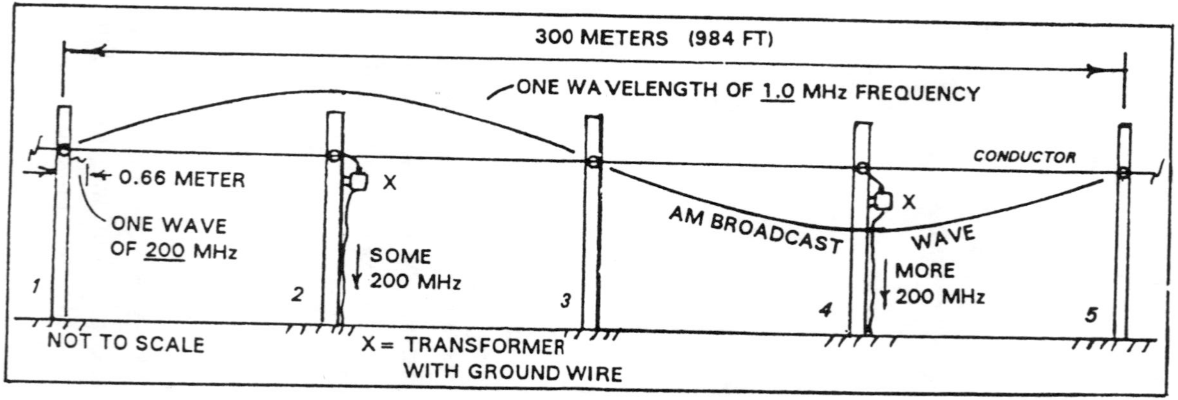

Fig. 7-15. Comparison 1.0 MHz-Wave vs 200 MHz Wave on Distribution Line.

Propagation of high versus low frequencies is shown in Fig. 7-15. This might help you better understand the rapid locating methods described. From a single sparking source, of course, we have a huge bundle of noise frequencies emitted, but only two frequencies are shown in the figure, 1.0 MHz and 200 MHz. The noise is arriving from the left of the picture, and some of the high frequencies will be bypassed or "filtered out" at the transformer locations.

Fig. 7-16. TVI Pattern from Antenna Preamplifier Overload.

Most TV sets can handle a signal level of 3000 microvolts or more on a particular channel without overloading, but others are less tolerant. It is easy to exceed this 3000 μV where moderately strong signals are coming into the antenna and an amplifier is used in the lead-in.

Where a broad-band (all channel) preamplifier is used, the overload point of the amplifier is determined by the total amount of signal coming in, summing together all the channels received. Overloading (excessive signal) may actually cause an amplifier to go *into oscillation* and an interference pattern will result. This interference may completely cover the screen or occur in narrow or broad bands; it may be seen on some channels and not others. Further, an oscillating preamplifier can broadcast TVI to other television sets in the neighborhood.

Cross modulation is a variation of overloading, where excessive signal occurs within a single TV channel where its video carrier beats against its sound carrier. It can also occur between adjacent or non-adjacent channels where carriers of one channel beat against carriers of another.

The overloading may appear on the TV set as horizontal streaking or smearing, random grouping of interlace lines in a more-or-less venetian blind pattern. The horizontal blanking and sync pulse of one channel may appear in another channel, or the vertical blanking pulse of one channel may be seen in another. Also, two pictures (adjacent or non-adjacent) may be seen on the same channel.

Fig. 7-17. Sync and Blanking Pulses of Channel 11 Appearing in Channel 9 Picture.

Cross-mod overloading is seen above. Here we see the narrow *horizontal bar* that the girl from channel 9 is attempting to erase (!) is the *vertical* blanking pulse from channel 11.

The very wide vertical bars in the figure represent the horizontal blanking and sync pulses.

When overloading is present, one can determine whether it is in the TV set or the preamplifier by inserting about 20 dB of attenuation between the amplifier and the TV set. If the distortion clears up, the set was overloading. If it does not, overloading of the preamp is indicated. If an attenuator is not available, and for a quick check to see if the overloading disappears, remove one wire from the receiver twin-lead connection for the test. This will reduce the level into the set. With high-band reception, both connectors can be removed and held slightly away from the terminals; the signal will jump across an air space on the order of 1/8". Note that higher channels jump the air space much more readily than the lower.

As pointed out earlier, a symptom of ghosting is a vertical pattern on the screen consisting of displaced horizontal sync and blanking pulses.

If the interfering TV pattern drifts back and forth, it is a symptom of overload; if it is steady and remains in the same location from day to day, it is a symptom of ghosting.

Preliminary oscilloscope control settings for the Leader 315A and the Hitachi 509 scopes are shown in the Appendix.

These scopes are not longer being manufactured, but recently have been available on the used market.

The Leader Model 315A should be your oscilloscope of first choice for power-line noise-source locating.

Chapter Eight

POWER LINE INTERFERENCE–SOUNDS, PATTERNS AND MYTHS

Introduction

This chapter describes and challenges certain myths about power-line noise which include interference to radio amateur and AM radio, to television, and to other services throughout the electromagnetic spectrum.

Several observed oscilloscope patterns from "sparking" RFI show that there is no special and characteristic sound for power-line noise, contrary to a common myth. They also show that the severity of radio *and* television interference is related not only to the *level* of the interference, but also to the *characteristics* of the noise *patterns*.

Another myth dispelled is that power-line noise results from harmonics of the power-line frequency. This myth is dispelled when it is shown that the sparking pattern for direct current (DC) sparking maintains itself throughout the VHF and the UHF frequency spectrums, even though no 60-Hz fundamental frequencies (or harmonics of 60-Hz) are present.

New insights were obtained by comparing, at a fixed voltage, various sparking-gap sizes with their effects, that is, with the resulting oscilloscope patterns, radio noises, and TV picture interference.

Pattern Correlations with Sound and Sight

A "power-line noise" generator with an adjustable spark gap simulates numerous RFI patterns seen in the field as radiated from power lines. For comparison, sparking patterns are shown in Table 8-1, page 8-4, as resulting from DC power, as well as from AC power.

The observations were made using the set-up as shown in the block diagram of Fig. 8-1.

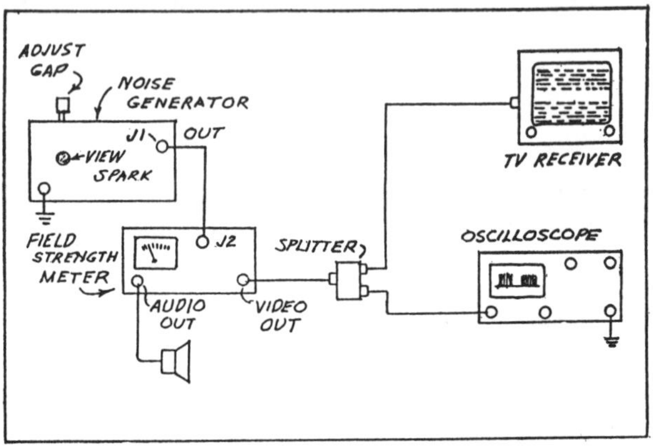

Fig. 8-1. Block Diagram Set Up for Sparking Effects Observations.

The variable spark gap in the noise generator is an automobile-ignition point-contact set, adjustable from outside the metal enclosure of the generator by means of an insulated tool. The receiver used is a field strength meter (FSM), Sadelco Model FS-3C. The oscilloscope is a Leader Model 315A.

Power for the noise generator is supplied by a 1.5-kV transformer, except for the final step shown in the Table, where a pure 3-kV DC voltage (derived from an oscilloscope power supply) is applied. The patterns shown are some representative sparking groups with different group spacings and different "spike" densities. The sounds and sights observed with these patterns closely approach many of the sounds heard, and the TV picture interference observed, in the field.

The variations of effects with spark gap size variations at a fixed voltage, are also shown in Table 8-1.

The remarks shown in columns 4, 5, and 6 of Table 8-1 are based upon the subjective observations of the author. Remarkable differences in radio-noise sounds and TV picture disturbance from the different, but typical, power-line sparking patterns were observed.

The character of the interference-sounds heard will vary with the characteristics of the radio receiver used. The amplified output of a field strength meter with no squelch or no special threshold provides a good comparison of noise sources. If observations are made with a receiver "riding on its squelching threshold," a crackling, scratchy sound may be heard, but it is not the true power-line sparking-noise sound.

Note that in the patterns shown, a dotted line for a spike, or portion thereof, indicates rapid intermittency.

TABLE 8-1. PATTERN CORRELATIONS WITH SOUND AND SIGHT.

RF RADIATION PATTERN DETECTED 1	REMARKS 2	APPROX GAP SIZE 3	"NOISE-SOUND" 4	RI 5	TVI 6
Reference 60 Hz 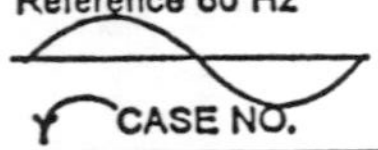 CASE NO.	60-Hz Sine Wave	-----	-----	-----	-----
#1 -----	Gap open, no sparking.	0.70mm	-----	-----	-----
#2 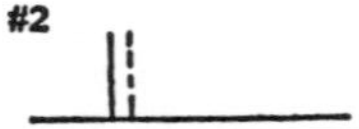	RF noise detection on positive half-cycle only.	0.46mm (.018")	Crackling sound. 60-Hz component.	Quite noticeable.	Noticeable, but picture watchable.
#3 	Detected RF noise on both half-cycles.	0.45mm	Sharp crackling sound with definite 120-Hz component.	Quite objectionable.	Same as above.
#4 	Jumping rapidly between patterns 2 & 3.	0.44mm	Very rough, crackling. Jumping between 60 and 120-Hz sound.	Very objectionable.	Objection-able.

#5 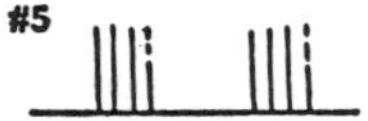	Fairly solid pulse groups on both half-cycles.	0.38mm	Very rough, bacon-frying noise. Definite 120-Hz component.	Very objectionable.	Quite objection-able.
#6 	Pulse groups becoming dense.	0.36mm	Smoothing out, but 120-Hz component still heard.	Objectionable.	Very ob-jectionable.
#7 	Pulse groups becoming more dense.	0.25mm	Same, but with rushing sound.	Objectionable.	Severe.
#8 	Transiting quickly thru patterns 2, 3, 4, 5, 6, 7, and reverse.		Sounds like whistling through one's teeth.	Very objectionable.	Severe.
#9 	Very dense pulses, closely-spaced groups.	0.13mm	Quite smooth.120 Hz. component weak. Sounds like high background noise, rushing.	Less objectionable than (7).	Severe.
#10 -----	Gap closed. No sparking.	-----	-----	-----	-----
#11 	DC sparking.	0.56mm	Smooth. rushing, and steady. Slight whistling.	Less objectionable than (7).	Severe. Covers screen.

Pattern Importance to Noise Locating

The pattern retention throughout a broad frequency spectrum is illustrated, where Fig. 8-2 shows a 50 MHz pattern at (A). At (B) it shows substantially the same pattern at 200 MHz, viewing the same sparking source, and at (C) it views the same source at the much higher frequency of 500 MHz. This demonstrates the pattern retention with frequency increase which makes our rapid locating system possible.

(A)
50 MHz

(B)
200 MHz

(C)
500 MHz

(60 HZ)
Ref.

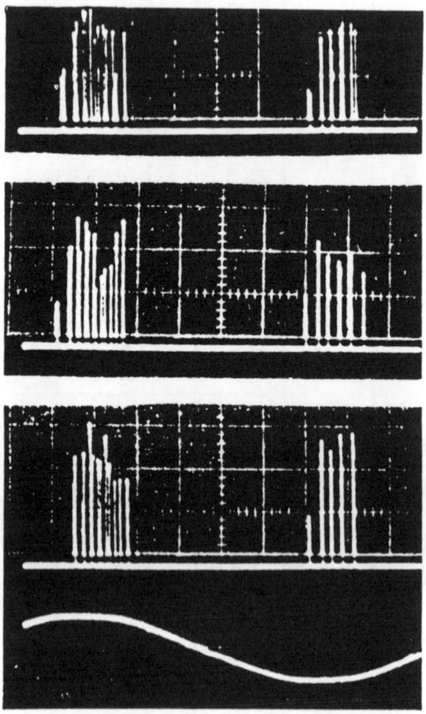

Fig. 8-2. Same-Source Pattern Observed at 50, 200, and 500 MHz.

At the 500 MHz frequency the noise level is lower than at 50 MHz, and the oscilloscope is adjusted to provide the same pattern height as shown in (A), for the sake of *pattern* comparison.

Myths and Misconceptions

Certain myths and misconceptions about power lines and RFI are listed here.

MYTH 1. The harmonics of 60 Hz (or 50 Hz if that is your power-line frequency) are responsible for the very wide frequency spectrum of power-line RFI.

No. It is a very common belief that the wide frequency RFI range generated by a sparking source results from harmonics of the power-line frequency. However, the author's study of sparking patterns on the oscilloscope in the field and in the laboratory indicates that this is not true because of these two observations:

(1) With his laboratory "power-line" noise generator," the author was able to generate pulse groups of various spike densities and group spacings. He found that pattern #2 of Table 8-1 had substantially the same frequency spectrum as patterns #3 through #9, and he submits that the frequency spectrum is dependent only on the *very sharp rise-time* of the individual spikes in the groups. These very sharp disruptive spikes, each with a rise-time of about 2 nanoseconds (2-billionths of a second)[2], seriously disturb the electromagnetic environment–and their frequency range is not dependent upon the harmonics of the power-line frequency. Further it is not dependent on the number of groups in one cycle nor on the number of spikes in a group.

(2) As shown in pattern #11 of Table 8-1, *DC sparking* spikes fill the width of the oscilloscope screen, and there is no grouping in the pattern as with AC. The pattern at step

#11 looks very much like pattern #9, except that there are no groups, only a continuous field of spikes.

> An air gap will break down and spark whether the voltage across it is AC or DC. The RFI-generating spikes will be present in either case, and the harmonics of 60 Hz do not enter into determining the RF noise frequencies generated from sparking.

MYTH 2. DC-power lines will not cause RI or TVI.

False. It is possible. See Fig. 8-4.

It is obvious that believers in Myth #1 would not expect DC power to radiate noise at television frequencies, because of the lack of harmonics from DC. A number of power engineers have observed that RF radiation does take place, and that where DC sparking is present, the sparking source will radiate at frequencies up to 800 MHz [4].

It is shown here for the first time in Figs. 8-3 and 8-4, however, that the oscilloscope patterns for DC sparking are closely related to the patterns for AC sparking, and differ only in the lack of "spike grouping." Also shown are TVI observations corresponding with the scope patterns observed. The spectrum for this DC sparking exceeds 0.5 to 800 MHz (the spectrum used by the author for his rapid locating system). These are laboratory observations.

Fig. 8-3. DC Sparking as Seen on Oscilloscope at 76 MHz.

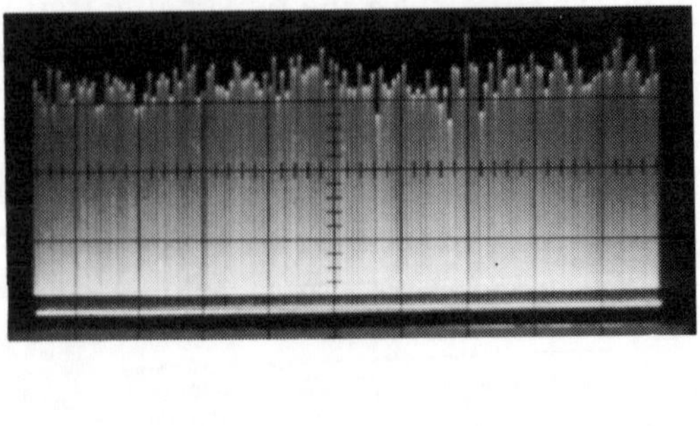

Fig. 8-4. DC Sparking as Seen on TV Set at 76 MHz.

Sparking occurring on an operating EHV (Extra High Voltage) DC line was shown occurring by Pakala and Chartier in 1970 [4].

MYTH 3. Power-line interference has a characteristic sound. *(You may have heard a statement similar to "I can tell it's power-line interference by the sound. To me, it sounds like a crackly or popping noise.")*

> ***No.*** Sometimes it sounds as described above, but frequently not. See Table 8-1. Also the noise you hear may not be a source on the power line at all, but a connected device reradiating noise from the line.

MYTH 4. The very best way to locate power-line sources is with a spectrum analyzer.

> ***No.*** It is an inefficient method. Some people maintain that the only way to locate power-line electrical sources is to make observations "in the spectral domain rather than the temporal domain." This sounds very scientific but, in fact, is not helpful in locating a power-line source-containing structure. An example follows:

Assume you are near the residence of a complainant who has "bad interference" on TV channel 2, (54 MHz), but on no other channels. The house is on a corner next to a 4-way pole and multiple-sources are present. Observing with your spectrum analyzer, you can drive in any of four directions from the house and the spectral display might be markedly different at two poles away in each direction (because of propagation effects...see Chapter 3).

For comparison, at the same starting point with your time-display scope, you see three separate and different patterns (on your scope at the same time) at 54 MHz, with one of the patterns being much higher than the others and somewhat different in shape than the others, so you travel in the direction of the highest pattern at the highest frequency observed and then locate a source structure by following the same pattern to the source pole.

In looking for *signal (single-frequency) interference* at a discrete frequency, a spectrum analyzer may be used to find one such source. But discrete signal interference is not what we are looking for in our search for *power-line* or other *electrical* sources. Rather we are looking for broad-banded electrical sparking-noise, traveling, at its start, in a large bundle of frequencies.

MYTH 5. "If a source is present on the power line, an ultrasonic detector will detect it; if I get no ultrasound pickup, there is no power-line noise."

<u>***No.***</u> Even the most sensitive ultrasonic locator will not pick up sparking if its source is hidden from view of the locator. This does not necessarily have to be a direct "line-of-sight" view as some nearby reflected ultrasonic radiation <u>may</u> be picked up. There is a real place for that "most sensitive ultrasonic locator," however, as described in *Chapter 4*.

MYTH 6. Lamp dimmers are the worst invention of all for generation of noise (and reradiation from power lines).

<u>No.</u> They do not compare with doorbell transformers. The dimmers checked by the author did not radiate above 8 MHz, whereas doorbell transformer radiation exceeded 0.5-800 MHz. The transformer thermostats generate broad-band noisy sparks, and dimmer switches do not.

MYTH 7. Power-line interference is caused by microsparks, not by arcs.

<u>This is a matter of definition.</u> An arc can be a power arc or a small arc. This small arc is variously called a spark, a transient arc, or a microspark.

MYTH 8. Corona is <u>the</u> cause of power-line interference.

<u>No.</u> Many people, even some power company personnel, believe that corona is *the cause of power-line interference, when actually, at least 95% of the sources causing complaints from power lines are found to be sparking sources.*

MYTH 9. If both transmission lines and distribution lines are present, you should be more concerned about the high-voltage transmission lines because they <u>really</u> cause noise.

<u>No.</u> Overall, many more sources resulting in complaints are found on distribution lines than on transmission lines. However, you *will* find very noisy transmission lines, especially these are on wood poles and have loose hardware.

MYTH 10. Transformers on distribution power poles are a most likely source of power-line interference.

<u>No.</u> Pole transformers only <u>rarely</u> cause *RFI.* See Chapter 4, page 4-15.

MYTH 11. In looking for power-line interference, if the complaint is on a particular TV channel, say 5, be sure to continually observe on channel 5.

No. This will lead you astray. Because of the very broad-frequency spread of electrical sparking radiation and its propagation characteristics, when looking for sparking sources you should observe at the highest frequency at the highest level where you can see the noise pattern of concern on your scope.

MYTH 12. Power-line interference always disappears in rain.

No. Sparking power-line interference usually disappears in rain, but not when a source is in a protected location or in an enclosed device such as in a lightning arrester.

MYTH 13. "There is one bad source out there causing the trouble; if I locate it, that will solve the problem."

No. Multiple sources are usually present, and are actually more the rule than the exception. A report you may hear is, "That noise source out there is getting really bad!" ...when actually another source or two may be joining in.

MYTH 14. On either transmission lines or distribution lines, the source is likely to be a "bad" insulator.

No. It is more likely hardware "connection" problems, or connections at or between insulators.

MYTH 15. The power-line interference increases as load on the line increases.

Not likely. When the load increases the voltage may decrease significantly, and some of the sparking sources may in fact disappear, and the interference level decreases.

MYTH 16. You can trace a noise to its source by its sound.

Sometimes. But it is very difficult if two or three sources with only slightly different sounds are present (A scope is of great help in discriminating between noises because different noises present different patterns.)

MYTH 17. Correction is not a problem; it's the locating that takes the time.

No. An experienced investigator stated this a number of years ago years ago. In the 1940's efficient locating equipment and methods were not available, and this explains the statement.

MYTH 18. Once the source pole has been identified, the problem is likely to be very obvious from ground observation.

No. Very often a source cannot be identified by visual observation from the ground because there usually are many possible source locations on the pole. These locations may look OK, but a connection that looks tight could very well be sparking. The *general area* of the source(s) on the structure can often be narrowed down, however, with a 300 to 800-MHz UHF observation from the ground. (For pinpointing from the ground, the higher the frequency that can be used the better.)

MYTH 19. If every thing is tight and is kept tight, we will have no noise problems.

No. Connections may feel tight to the maintenance person, but sometimes they are only tight because of corrosion in the connections.

MYTH 20. Because of their low voltages (120/240 volts) service lines rarely cause radiated noise problems.

No. Service lines *frequently* radiate noise originating in a home, from faulty door-bell transformers and other devices operated at service-line voltages. In this book, of course, the author defines this as *reradiation*, because here it is being observed on the line rather than at the actual source location

MYTH 21. Generally, power lines cause AM-radio interference, but little TVI.

No. This, of course, is not true, but the statement is found in the literature. If this were true, the author would not have written this book, because serious TVI from power lines prompted the research described.

MYTH 22. ***If it sounds like 60 Hz, it is power-line interference.***

Could be. However, you might also be hearing the 59.94 Hz video carrier of a TV transmission, or it might be the dimmer for an incandescent lamp connected indirectly to the overhead power line.

MYTH 23. ***"Trash" on the conductor causes very local noise only.***

No. As an example, on a 115-kV transmission line conductor, noise was observed at 200 MHz at a structure one mile from the source, as a result of a 2-ft piece of wire hanging on the conductor. [Propagation of noise at the high frequency of 200 MHz for this distance would be highly unlikely on a distribution line with its frequent pole transformers, its short spans and its "short" insulators.]

MYTH 24. ***The frequency spectrum of the spark depends on the size of the spark gap.***

No. The author has not found this to be true, in the range observed, 54 to 800 MHz, and with small gap sizes of up to 0.1 inches (.25 mm). This may not hold for larger gaps, although the author has found that all external sparks can be detected up to at least 500 MHz, *from ground level* at the source structure.

MYTH 25. ***Clamp-top post-type insulators are the answer to your noise problems.***

No. Not necessarily, unless pivot bolts are snug in their trunion assemblies and associated hardware is tight.

MYTH 26. ***Distribution lines have ground wires on every pole.***

Sometimes, on almost every pole, at other times, only at transformer poles or capacitor-banks, and this depends on the particular power company.

MYTH 27. ***I don't need better equipment: I can always locate the sources with my AM radio.***

No. Sometimes you can, if only a single source is present, but usually not pinpointed with certainty. With experience you do get better at guessing the right pole, with only a clue from the AM radio.

MYTH 28. ***Overhead power lines are the primary source of radio noise in the MF frequency range (3 to 30 MHz).***

No. Overall, when both distribution and transmission lines in urban and rural areas are considered, power lines contain less than half of the noise sources causing complaints. In many cases, however, the noise conducted into power lines from devices in homes and other buildings will reradiate from the power lines, and the power lines then may be blamed for these sources. This applies as well to VHF frequencies (30 to 300 MHz), except that the VHF frequencies are much weaker at the sparking source and do not propagate as far along the power line as the MF frequencies; their radiations are not as widespread and would not generate as many complaints.

MYTH 29. ***Power-line noise-source locating efforts are invariably hampered by the intermittent and erratic behavior of such noise sources.***

No. This is not true if the RFI investigator is properly equipped and investigates the power-line complaints when the complainant says the source is "on," or is likely to be on. For example, if a radio amateur operator says the noise always comes on about 3 pm (in fair weather), the investigation should be made after 3 pm in fair weather. Further, fairly steady sources are observed much of the time.

MYTH 30. Essential to noise generation is a grounded metal object being in close proximity to a power line.

No. Two objects in close proximity and in the electric field of a power line (but not *on* the power line) may spark between each other. These objects need not necessarily both be metal, but for significant noise radiation to occur, one object must be metal to provide a "broadcasting antenna" effect. No grounding of either "gap electrode" is required.

MYTH 31. "All that crackling and buzzing we hear near this substation means the station is causing serious radio and television interference."

No. The noise we hear "with our bare ears," is probably not serious, and is probably corona. If this substation is operating at 230 kV or below, the RFI from this corona will probably not be observed beyond 300 ft (100 meters) from the substation.

MYTH 32. (Actually we have four myths here).

32A. Doorbell RFI can be cured by disconnecting the bell (or chime).

No. Disconnecting the bell from the transformer will not kill the RFI because the transformer is always heating to some extent, as it is (almost always) *directly-wired into the service entrance panel.*

32B. Doorbell RFI can be identified because its RFI noise is ON for one second, OFF for seven, and then repeats.

No. The on-off pattern cannot be specified as in this statement. Many different on-off patterns are observed in the field.

32C. Doorbell RFI is caused by a staple shorting the doorbell wire.

Probably not. This will also cause the doorbell to ring, continuously or intermittently, depending on how solidly the staple shorts out the wires.

***32D.* Doorbell transformer RFI is caused by a sticking doorbell button.**

No. This will cause the bell to ring and continue to ring. You will be aware of it, if this is the problem.

See Fig. 5-21, Chapter 5, for the doorbell transformer and bell circuit.

MYTH 33. The higher frequencies travel much shorter distances than do the low.

Not exactly. In your locating experience this may appear to be true. Here, however, in locating power-line noise, you are dealing with *propagation* via the power line, for the most part, rather than *radiation* through the air directly from a source. In the power-line propagation mode the higher frequencies are "filtered out," to put it simply, as this propagation along the line takes place.

> Also: An important fact to remember is that the higher frequencies from a spark, at the source, are starting their travels at a much weaker level than the low frequencies. See graph in Fig. 17, *Chapter 1.*

If you are a radio amateur, you are often dealing with signal propagation via the global atmosphere; here the ionosphere and other effects refract the lower frequencies. These effects are frequency selective and in general permit the lower frequencies to travel much further than the high, often with these lower frequencies more-or-less following the surface of the earth.

Think of the moon-bounce experiments by radio amateurs where very high-frequency signals (150 MHz) with earth, in a single *reflection*, traversing nearly a half-million miles.

MYTH 34. ***Power line noise sounds like bacon frying.***

Not necessarily. Sometimes it has a sharp staccato sound, a buzzing sound, a sound like bacon frying, a whistling, or a rushing sound. Almost always (if you listen closely), it has a 60-Hz or 120-Hz character. Try not to confuse this 60-Hz character, however, with the 60-Hz (approximate) repetition rate of a TV synch signal, or confuse it with the ratty 60-Hz sound of a dimmer switch or touch-switch lamp. Dimmer switch reradiation, you should note, will generally propagate only for a few poles from the source.

MYTH 35. ***Noise generation results from the arcing action of objects in close proximity to a power line which creates relatively smooth, square waves composed of many high RF frequencies.***

No. In actual fact the noise generation resulting on an AC power line is caused by the very sharp RF spikes caused by sparking. These occur in more-or-less rectangular groups as shown in Table 8-1. It is the very sharp individual noise spikes that generate an extremely broad-band frequency-bundle of frequencies from below 1-MHz to above 1000-MHz.

From an AC power line it is the *repetition rate* of the somewhat rectangular groups (of spikes) that imparts the 60-Hz or 120-Hz *audio character* to the buzzing or crackling noise heard in a radio.

Chapter Nine

STORIES FROM THE FIELD

Introduction

The stories in this chapter are based on the personal experiences of the author *(mol)*, of Mike Martin, Potomac Electric Power Company (PEPCO), Washington, D.C; of Jim Hope, Public Service Company of Oklahoma; of Jack Talmadge, of Radar Engineers; of Dave Creighton, formerly of Pacific Power; and of Alan Robinson, of Oregon State University.

Three stories were told in previous chapters.

STORY 4

THE PAD MOUNT TRANSFORMER HAS TO BE REPLACED

by Mike Martin

I received a call from a power company in my area with an "arcing problem." This was reported by a customer to be that big "green thing of yours," meaning a pad-mount transformer (a transformer on a concrete slab in an above-ground vault).

The customer, using his broadcast band hand-held portable radio, followed the noise out of his house to the vault, and got a very high level of noise there. He called the utility and they investigated and agreed it must be a faulty transformer.

They shut if off and the noise stopped. Well, that proved that, so they thought! They replaced the transformer, and when they energized the new transformer, they heard the same noise as with the old one. They also heard "Whose idea was this?!"

After I arrived, I was able to locate the noise source a short distance from the vault inside a next-door neighbor's home.

The source was a recessed-housing ceiling lamp that was thermally-protected, with its thermostat sparking. This noise would radiate throughout the house and conduct itself to the transformer. It would then reradiate and propagate back to the other houses connected to that transformer.

The ceiling lamp was switched off, the noise immediately disappeared, and the complaint was resolved.

This one mistake made by the power company would have easily covered the cost of calling someone who specializes in locating interference sources.

STORY 5

THE RELUCTANT FORK-LIFT DRIVER

by Jim Hope

It was a cold November day when I got a call from a CATV owner in one of the North Central states. He was complaining about having electrical interference from the power company that was ruining his signal from a television station in Shilo. He stated that the power company had been notified and that they had been out several times to work on the lines during the past 18 months, but he still had the same problem that kept showing up at irregular intervals during the day.

His opinion of the power company was anything but great. He asked if I could come up and see if I could locate the interference so he could get the power company to fix it. He was not only catching heck from all his CATV customers but was also catching flack from a television station because viewers were turning away from their station, and advertisers were complaining that they were losing the promised number of viewers.

I agreed to go up and locate the interference, so I booked a flight for the next week, reserved a room in a hotel, and reserved a rental car. I made a call to the local power company and made arrangements to meet with them to help work the problem out.

When I arrived in Shilo, I loaded my equipment in the rental van, met with the technician from the power company, and we headed north. Upon arriving at the head-end site of the CATV company, we were shown the interference on their monitor. I agreed that they had electrical interference and that I could locate it without a problem. As I got out my equipment and began looking, I was able to determine that the problem was coming from due south. This was the direction that the CATV antennas were having to look for the television station concerned. As I progressed to the south, the level of interference became stronger with each passing block.

Soon I was standing in the middle of a lumber yard that was 1/2 mile south of the CATV tower. My equipment was pointing directly at a large yellow forklift. The operator asked what the heck I was doing pointing that TV antenna at his fork lift. I explained what I was doing and asked him to shut the machine off.

He told me that I was crazy. He said they had some kind of problem with it, and that it was so hard to start that they let it run all day and only shut if off at night. When I explained the importance of finding a solution for the CATV company he finally agreed. When he shut it off, my equipment got quiet.

I went straight to the CATV head-end and they were *jumping up and down with joy* (see Fig. 9-1). Their first question was how did I get a line crew to repair the problem so quickly? When I explained what and where the interference was, they were not only surprised but were embarrassed that they had given the power company such a hard time. I assured them that I would explain the problem to the power company and the television station engineer.

Cases like the one above are typical and are another reason that it is not possible to think that all electrical interference is generated by power company lines. The amount of money that it cost to send a line crew out to try to solve a problem is so much today that it is not the feasible solution to the problem. No matter how much money you throw at a problem, the customer is only going to be satisfied if you solve the problem. The moral to the story is, if you don't have an RFI expert, call one.

Fig. 9-1. Jumping with Joy. [Story 5]

[Equipment Mr. Hope used to locate the fork-lift truck: 237 VHF-UHF locator with 500-MHz antenna with extendable antenna dipoles for VHF use. At the head-end site, he started the locating procedure using a low VHF TV-band frequency near channel 2. This equipment has now been superseded by the Radar Engineers Models 239 and 240, which cover the required frequencies...ed.]

A DEMONSTRATION

As an introduction to Story 6, a demonstration you can easily make about radiation and propagation is as follows: Take your transistor radio and, with it tuned to a weak station on the AM dial, hold it close to a vertical ground wire on a distribution line pole about 4 ft. above ground. You will notice the volume of the music you hear increases considerably as you approach the ground wire.

This indicates that the power line is receiving AM radio signals; it is propagating them; and it is reradiating them. [This reradiation is somewhat frequency-selective, so trying it at another spot on the AM dial may improve your demonstration.]

STORY 6

TALK SHOW INTERFERENCE

(by mol)

I was called to investigate an AM-radio complaint from a customer living in a rural area near Olympia, WA. I will call him Gordon. He was both insistent and persistent. That is, insistent that it was power-line noise, and persistent in his phone calls to the Bonneville Power Administration, the power transmission company, and to Puget Sound Energy, the local distribution company.

Crews from both power companies had been out to check their lines in the area and had not found any noise on their lines. On my first visit I also found there was no power line noise.

It took a fairly long time to gather some useful clues. Gordon said that the noise severely disturbed his favorite talk show station–especially on Saturday mornings (when he was off work). He indicated it was intermittent at other times. His radio was located on a chest in the living room, close to an inside wall, and he said that was where he got the best reception.

The station in question was KXL at some distance south (100 mi) in Portland, OR.

I did conclude on my first visit that the location he had chosen for his radio was a good one because the power line, acting as a receiving antenna, was picking up broadcast signals, propagating them via the service line into the house wiring, and into the wiring behind the radio there it was *reradiating* the AM-talk shows of KXL into Gordon's radio.

My next step was to locate the source of the noise coming in on the service line by checking it on Saturday when it was said to be on. See Fig. 9-2 for locations.

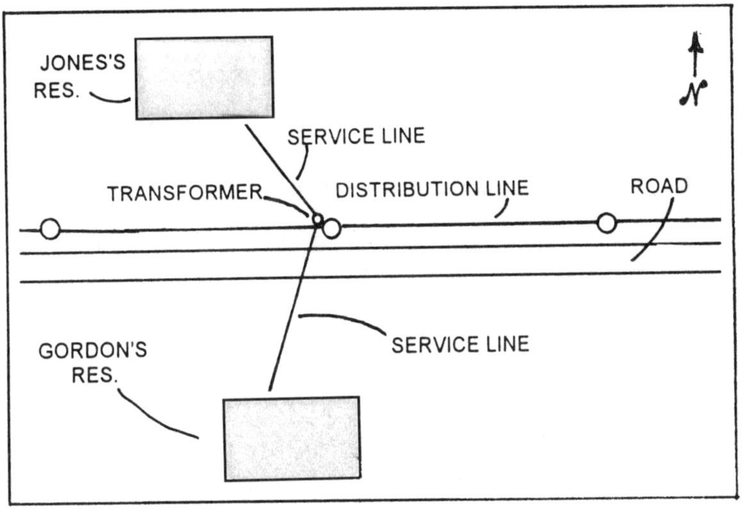

Fig. 9-2. Sparking Noise from Neighbor's Home.

The noise did not appear on my first Saturday visit, even though it was a fair day. The next Saturday, though, I found noise coming in from the north, from the direction of the distribution line across the road, and beyond the distribution line, from a neighbor Mr. Jones's house. At this residence (which was connected to the same pole transformer as was Gordon's home), I found a very noisy electric blanket with a badly sparking thermostat. Unplugging the blanket immediately stopped the noise. As a test we plugged it in again, and the noise came back, and so did Mr. Jones's bad TVI. The blanket was then, of course, left unplugged. By identifying and disconnecting the noisy blanket, the noise pickup was eliminated, and Gordon and Mr. Jones then had good reception on radio and TV.

While the distribution line and its service lines had provided the means of Gordon's good signal pickup, they also provided the reradiated noise from his neighbor's house!

STORY 7

GET YOUR SPEED UP WITH MADE-FOR-THE-JOB EQUIPMENT

by Mike Martin

As an RFI investigator/consultant, I'm often called upon to locate the source of television interference thought to be from a power line. As with the following case, often the investigator's lack of success is due to the inadequate equipment used in trying to locate the problem.

I was called upon to locate the source of a TVI problem that seemed to be affecting several customers in a two-block area.

The TV channels affected were mostly 4 and 5, but to some extent, channel 7. The power company, using a portable AM broadcast-band receiver, had supposedly "located" the source several times over the past couple of years, but they seemed to still have TVI.

As it turned out, the area was affected by several sources of power-line noise, and the AM broadcast band receiver just wasn't the right instrument to verify which noise sources were bothering the customers (the complainants) and which were not. With my Vehicular Locating System (VLS), I was able to start at the customer's home and locate and correct just the sources affecting the TV. The complainant had TVI on channels 4-5, so I started in front of his home just above the channel 5 frequency -- just far enough above it so that I could not see, on the scope, the vertical pulse of channel 5.

With the use of (1) the Radar Engineers 239 VLS receiver, (2) the directional antenna and (3) the Hitachi V509 oscilloscope I was able to figure out not only the direction of all the noise sources receivable at the customer's home, but also evaluate the noises in order of amplitude. With this information, and with my handy line crew, I was able to make repairs to the noise source with the highest level and then find that some of the lower level sources weren't affecting the customer's TV.

It's important to note that not all noise sources in an area will be affecting the complainant, and that not all TV's in an area will be affected by the same source. It's important to be able to pick out each power-line noise source individually.

Equipment other than RE Model 239, such as a broadcast receiver designed for receiving *signals* from all directions, will also receive *noise* from all directions. The VLS can concentrate on receiving noise(s) from one direction at a time, compare noise amplitudes, and allow the investigator to concentrate on the highest noise in that particular direction.

After investing in the VLS system, the power company referred to has reduced its correction time by 90%. They also no longer have to make unnecessary repairs on equipment not interfering; this means they can respond to more complaints in less time.

STORY 8

THE ELECTRIC FIELD MAKES A STATEMENT

(by mol)

A source of non-power line electrical noise had a cable TV company, a municipal power company, and a sign company baffled. This is a good illustration of the importance of the electric field as a factor in noise generation. It also illustrates an unusual correction procedure.

A CATV company called me regarding a neon sign causing television interference on channels 2 and 6. These channels were being received by their antennas at their head-end site from TV stations in Portland, OR, about 20 miles away.

The cable company experienced the interference only when a large supermarket neon sign was turned on. Their antennas looked directly toward this super-market about 450 ft away. They had concluded that the neon sign was the problem, but all efforts by the sign company to correct it had failed.

They furnished the following information, which were good clues:

1. The noise would be on during fair days and would continue on during short, light-rainy periods. It would stop during steady, heavy rain.

2. The noise would stay on if either the left or the right half of the sign was turned off.

3. The sign company had checked connections to the neon tubes, and to all of the five transformers used in the sign, and said they were OK. Suspecting bad transformers, their next step was to be replacing these five transformers, four 12 kV and one 9 kV, with much more expensive types.

I did not suspect a defective transformer, because internal sparking in transformers within well-shielded and grounded enclosures, as these were, would radiate little noise at the TV frequencies. Also, internal sparks here would not be shorted out by rain.

Another possibility would be arcing in the indoor service panel breaker, but I ruled this out, because if such a breaker were arcing, it would *not* be shorted out by rain.

I was advised that the neon sign was at the present turned off, and was being left off, as a public service gesture by the store manager who was anxious to find out what the trouble was, as was the sign company.

I advised the cable TV company that I would come over to look at the problem on the first fair day I could get away. I would not want to be there if the noise were suppressed by wet weather. This was per my rule to try to locate interference only when it is *on*. That sounds logical enough but is often attempted otherwise!

In a couple of days then, I met Jeff from the cable company at the supermarket. Upon arriving, we had the store manager turn on the sign. Sure enough, while I was parked in front of the store, I could immediately observe the interference at 200 MHz on my oscilloscope. The sign is shown in Fig. 9-3.

Fig. 9-3. Neon Sign Containing Several Sources Radiating TVI.

Jeff had arrived driving his service bucket truck, and with my hot-stick sniffer he went up in the bucket and started scanning the sign, with the 110 MHz rod-antenna of the sniffer. By this means he detected peaks of radiation near the letters "h" and "a."

With this observation, we were getting close to the problem, and it was time to call the sign company for help from their crew.

The store manager then called the sign company in Portland, and they said they would have a man there within an hour.

When Dave from the sign company arrived with his ladder truck, we asked him to take the hot-stick sniffer and scan the sign to see for himself that the noise was coming from the sign. He did this and also noticed that when he pushed the sheet-metal frame surrounding the letter "h" against its backing raceway, it killed the noise heard in the sniffer. I could still observe it in my vehicle, however, but it was weaker. The same thing happened pushing in the "a;" this also partially killed the noise. He noticed, too, that when he touched these metal letter frames barehanded he felt a slight "tingle." These were two very good clues.

Before going any further, the construction of the sign should be explained. The complete sign was shown in Fig. 9-3. Neon tubes light up each individual letter, and these tubes protrude from the long sheet- metal rectangular sign-transformer housing, which encloses five sign-transformers, and provides a backing metal raceway for the neon letters of the sign. When turned on, the neon tubes shine through the front red plastic panels of the individual letters. One of the letters, "I," removed, is being held by Jeff in Fig. 9-4.

Fig. 9-4. Cable TV Engineer Jeff Hurd Holding "I" of Noisy Sign.

The thin sheet metal frames surrounding each letter look like over-size "cookie cutters," and these are attached to 5/8" plywood "cookies" with several wood screws for each cookie. The cookie cutters are called "chases" by the sign company people, and the cookies I will just call "plywood bases."

Now to explain the noise-source clues mentioned:

(1) The slight tingle felt by Dave indicated the chase was energized, that is charged, with a voltage, and, (2) sparking was

indicated because pushing in on the letters killed the noise and was obviously shorting out spark gaps.

After turning off the sign, Dave removed several screws holding the "h" and "a" chases to their plywood bases and then removed the chases. The sign was then turned on for a short test, and NO noise was heard. The chases were examined and two or more "burn" spots were found on the bottom edges of each of the letters removed, indicating sparking burns.

There were two possible solutions for the correction: firmly bond the chases to the raceway metal, or insulate them from the raceway. The overtime clock was running, so it was decided to demonstrate a new application of Scotch 33 electrical tape, and it was applied to the bottom edges of the two culprit letters in "Thriftway."

With these "taped" letters replaced, the sign was energized and all was quiet.

I intended this as a temporary cure and so advised the CATV people. Nevertheless, it may last for years, and the people concerned certainly will recognize the problem should it develop again.

The explanation of the sparking generated is, of course, that the electric field of the energized neon tubes caused the chases to be energized, charged with a sufficiently high voltage to spark over to the grounded metal raceway.

The reason the noise would disappear in rain is that the sign was exposed to the weather, and in a good rain water could run down behind the letters to short out the spark gaps. In a light drizzle there was apparently sufficient heat and air-circulation from the transformers and the neon tubes to keep the letters dry.

Expect the unexpected. The sign company people, the store people, and the cable company, were quite amazed at the cause and correction of this problem and, of course, were very pleased. Another example of the unexpected is given in STORY 9 and a good example of the electric field acting at a considerable distance is given in STORY 10.

STORY 9

CORRECTION BY HACK SAW

by Dave Creighton

My company asked me to investigate a ham radio complaint in the land where the buffalo roam, and I found, appropriately, that this involved cow horns on power lines. As you may not know, "cow horns" is a slang term used in some Western areas for the one-piece metal brace sometimes used to support wood crossarms, as shown in Fig. 9-5.

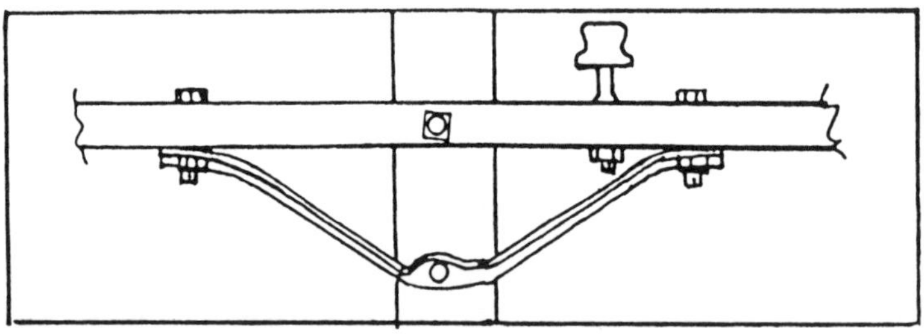

Fig. 9-5. The One-Piece "Cow-Horn" Brace.

I flew into this rather remote town on the company plane with my hotstick sniffer and my UHF locator, feeling I was well equipped to check out almost every possibility.

I arrived at the "cowboy's" ham shack and found him happy to see me...and hoping to finally get some relief from the power-line noise. I assured him that if it was on the power line, I would find it and the company would fix it.

He pointed in the direction that he thought the noise was located and towards a 12.5-kV distribution line. On the line I found with my

UHF locator that I could pick up the noise at 500 MHz; but also, I could pick up the noise at almost every pole at 500 MHz! What was going on here?

I suspected I had many sources, and I stopped to make a close visual check of one of the poles.

I noted a long insulator pin-bolt that appeared to be reaching below the crossarm and touching the cowhorn. The line crew was waiting for me to put them to work, so I sent them up to sniff out this pole with the 247 hot-stick sniffer, and they did find a high noise level at the location where the pin bolt was just barely touching the crossarm, as shown in Fig. 9-6.

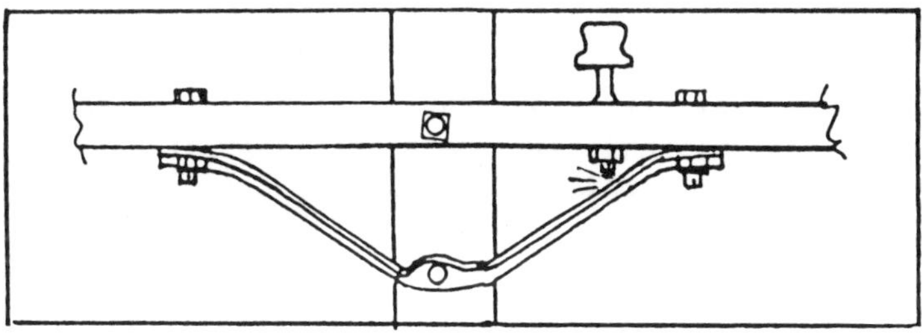

Fig. 9-6. Pin Bolt Sparking to Crossarm Brace.

The line crew then hack-sawed off the excess protruding-end of the pin-bolt, with the line energized, and this reduced, but did not eliminate, the high level of 500-MHz noise I had observed with my locator. Then we decided to check additional poles and see if this construction (or design?) error had been made at other poles.

Well, yes, it had, at only about a dozen poles, with several of them squawking all at one time. No wonder there were a few complaints among the townspeople! Even the sheriff was unhappy, but he didn't have the right equipment to corral the sources.

At any rate, all sources within reasonable range of the ham shack were corrected, and the complaint was resolved.

STORY 10

THE ELECTRIC FIELD REACHES OUT

by Jack Talmadge

I received a TVI complaint from a closed-cable TV company which had an installation in an apartment building near a Bonneville Power Administration (BPA) transmission line right-of-way.

The cable TV people thought the interference was from the BPA lines. I visited the apartment building and found that they did have electrical interference on their TV's which appeared to be power-line interference.

I connected my Radar Engineers 239 locator directly to the antenna used by the cable company at the apartment and listened to the noise between channels. There was no doubt about 60-Hz electrical interference coming in on their antennas, and probably it *was* power line.

Outside again, I used a hand-held 500-MHz Yagi antenna and vectored into the BPA lines near a 230-kV metal structure about a mile away, (see Fig. 9-7). As I got closer to the structure, it appeared that I was vectoring into a boat trailer directly under the 230-kV outer-phase conductor and near the transmission line metal tower.

I then used my 800-MHz Yagi antenna with the 239, and it was obvious that it was electrical noise coming from the boat on the trailer (also shown in Fig. 9-7, partly hidden by the tower).

Knowing that it was a 60-Hz noise, if there were a battery charger or heater in the boat, that would possibly explain it. However, with no 120-V AC supply at the boat this was not the case--and it must be a 230-kV electric-field problem.

So I took my hotstick sniffer and started checking out the boat; I quickly got to within 1/2" of the source.

Fig. 9-7. Boat on Trailer Under 230-kV Structure.

The sparking was occurring between a fish-net metal hoop to a metal tubing frame holding up the canvas boat-top. These were barely touching each other, and there was a sparking gap. (Fig. 9-8.)

I moved the fish net slightly and the noise stopped. This cleared up the TVI at the apartment building.

Here the electrical field from the overhead was charging the fish-net hoop and the metal frame of the canvas top to different potentials, and healthy sparking was taking place between the two.

You could blame either the boat on the trailer or the power line--both were required to generate the noise. You could also say the electric field from the outer phase of the 230-kV line overhead was really reaching for it!

Fig. 9-8. Sparking Location: Metal hoop to metal frame.

STORY 11

NOT TOO BAD A DEMONSTRATION

(by mol)

One Fall day I was putting on a workshop for a large utility in the eastern part of the country. In midmorning of one of our field days I was having a slight problem in finding noisy areas for demonstrations.

I had a power company RI/TVI investigator (whom I will call Jimmy) and two other company engineers riding with me in my VLS-equipped station wagon. Three other rigs including a line truck and crew followed. All told, I had about 12 people on this field excursion.

It was a beautiful day, and rather cool, and perhaps it was a mistake to schedule this outdoor class in the morning, because the lines were unusually quiet and possibly not dried out. However, Jimmy said he knew an area about 10 minutes away where a radio amateur was having a problem, and he would like to check it out.

We proceeded to the rather open residential area where the ham lived, and things were fairly quiet at his house. However, I could detect some noise to the east from my vehicle. While some of the people were visiting with the radio amateur I suggested I would take a few minutes to drive around a couple of blocks and would hopefully find a good source for a demonstration.

I didn't have to drive far because about two blocks away I was getting a strong noise on the upper TV channels. I could see a 12.5-kV road crossing about one block ahead. The dead-end structure north of the road looked like a possible source. I then brought my students to where I had parked.

After demonstrating what I could see on the scope screen in my vehicle, I let them try my hand-held locating equipment and suggested they check out the suspected structure.

Naturally, the structure was quiet, and I suggested it could be the dead-end structure on the other side of the street, so we crossed the street and ended up in a residential front yard.

The resident of the home came out to see what was going on in her front yard. She was very much interested and happy to let us proceed with the class.

One of the students then noticed that the noise was coming from the middle of the road rather than either end. With the 800-MHz locator he could point directly at the center of the span and at a piece of wire hanging on the conductor span which was obviously sparking badly. The source was found! See Fig. 9-9.

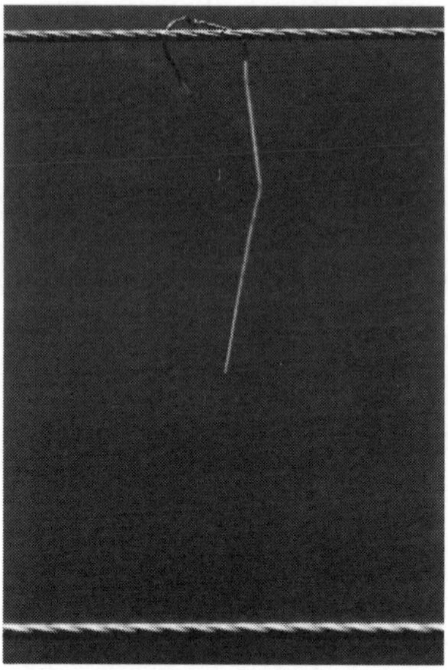

Fig.9-9.
A Sparking Source.

This was of interest to all, and they became believers in my locating method and equipment, but more was yet to come.

When the lineman brought the service-truck bucket up to the location of the span, I had everyone watch while first he checked it with the hot-stick sniffer so all could hear, and then had him remove the wire---and all locating equipment went quiet.

Then (to really make my day) I told them to watch what happened when he hung the piece of wire on the neutral conductor three feet below. This he did, and to their amazement, the noise came back on! (To be truthful, I was also somewhat surprised but did not let on.)

Here was another good example of the electric field reaching out. See Fig. 9-9. Here you can also see the *neutral conductor,* which I did not show in Chapter 1, Fig. 19, where I also used a portion of the same photo as an example.

I was accused, of course, of "planting" this wire (which was a broken and rusty coat hanger) and staging the event. It would be a good one to stage, but I had not.

I am sure the radio amateur of this story would have observed noise in this direction with his antenna being rotated toward the road crossing, because it was a high-intensity source, now corrected. Also, visiting a resident a couple of houses away we found him to be very appreciative that his bad radio noise was now cleared up. At any rate, all parties were pleased with the results, and I was pleased to get a rusty coat hanger as a souvenir.

STORY 12

THAT BIG TRANSFORMER'S CAUSING ALL THE NOISE

by Mike Martin

I was once contacted by a contractor who needed help in curing an RI problem. This problem had been active for almost five years. The complaint was made by an amateur radio operator who just happened to be a colonel on a military base.

The colonel had concluded that the problem had to be the transformer on the pole outside his house. Whenever he pointed his antenna in that direction, his noise level would peak.

Well, the power company did change that transformer, but the colonel did not stop complaining. It seems that noise wasn't coming from the obvious transformer after all.

I arrived and had the colonel demonstrate his noise locating with his rotatable-direction ham antenna; and sure enough, the noise peaked at a high level when his antenna was pointed right at that pole. Then I asked the colonel if I could take over his rotator controls. I found the noise level was almost four times as high when pointed 180° from the "obvious" pole. Now I was receiving a noise peak via the front of the antenna, and the colonel had been receiving a smaller peak via the back of the antenna.

As it seemed, the "obvious" transformer being south, when the colonel would look around for noise, he would look East, South, and West. He just neglected to go another 90° to North. I found the source to be a sparking loose tap at a lightning arrester on a pole to the North. The overhead crew tightened up the tap, and all is quiet now on base. By the way, the real source wasn't even on the base, as the colonel thought.

STORY 13

THIS TIME A CATV PROBLEM

by Mike Martin

I received a call from a customer who was very angry because of the "power-line noise" that was interfering with the reception on his television. After asking many questions I found the customer had the TVI while watching cable TV. The source was affecting the TV audio as well as the video.

The cable TV repairman had determined it must be in the power company equipment. I scheduled an appointment for the following day.

On my preliminary check of the area before entering the customer's home I did find power-line noises, one in front of the customer's home. I entered to find TVI on some, but not all, channels, and as the customer stated, it was affecting the audio as well as video. Because of the frequency of the channels affected and the level cable TV is operating at, I ruled out power-line noise as a source. Also, the video characteristics on the TV screen were not that of power-line noise.

I decided to check the cable tuner, because if it were in the *outside* cable system, I believe the cable company would have had many complaints. Also, another TV in the house on cable didn't have the problem. I disconnected the incoming cable from the cable TV tuner, connected the cable straight to the TV antenna input, and the problem was gone. I reconnected it the way it was, using the cable tuner, and the problem returned.

I phoned the cable company for the customer while at the location, and explained to them the procedure I used. Shortly, they sent a technician out to change the tuner, which corrected the problem, and the power company made another friend (the complainant).

STORY 14

THE COOPERATIVE CO-OP

by Alan H. Robinson, W4LKE

It all started about a year ago, in late March, 2001. That's when the power lines began to dry out after nearly six months of Corvallis, Oregon, rain. And from the end of March to the beginning of October I was able to make only a dozen contacts in the 3.5-30 MHz radio amateur band. The power-line interference at my home was intense; its noise level was over an S-9. By the end of September two things had happened: The rain came back and I had learned a great deal about the locating and correcting of power-line noise.

When I got back into ham radio during the summer of 2000, I hadn't planned on learning anything about power lines. Sure, we knew they ran past our house when we bought it five years earlier, but I never gave it a thought. The house has a nice view from up on a hill.

In October, 2000, I had set up my ham station and operated happily until that fateful day in April 2001 when it all stopped! I had made contacts in more than 50 countries when it all came to an end. When I turned on my receiver one sunny day I heard nothing but noise. So, I asked, how do you find it?

This led to my purchasing the ARRL RFI book, William Nelson's Interference Handbook, and finally Marv Loftness's AC Power Interference Manual. All three are helpful, but the real success came with Marv's book and his Rapid Source Locating Method.

At first I had tried using the AM radio in my car to locate the source. But if you're reading this book, you know that won't work.

After reading his book, I adopted Marv's method. I purchased a Radar Engineers (RE) Model 239 RFI receiver and an oscilloscope.

With this equipment I could look at the sparking pattern on the scope with my RE 239 hooked to my station antenna. I would then put the equipment in my car and look for the same sparking pattern from a nearby pole.

I used a 50/144 MHz whip antenna (MFJ-1728B) for proximity locating. Then, with a 440 MHz six-element yagi antenna I could locate the offending pole.

Fortunately for me, our local customer-owned utility has a very cooperative attitude. The district manager agreed to a plan for correction. I would find the noisy poles and they would fix them. Since then we have worked together and they were able to fix the sparking sources on the eleven poles I had identified.

It turns out that we have a 69 kV transmission line with a 12.5 kV underbuild distribution line running along the north edge of our property. The lines use dead-end bells for feeding tap-lines to pole transformers. The 69 kV has a mixture of pin insulators and suspension (bell) insulators, both causing noise.

Some of the sources required shutting down the 69 kV line in order to replace pole hardware supporting the distribution underbuild. Most of the sources were fixed by simply tightening the hardware on the wood poles, but some were caused by defective insulators and I was "awarded" several insulators to put on my fireplace mantel.

I decided to purchase a Radar Engineers Model 250 Parabolic Ultrasonic Locator to augment the Rapid Source Locating Method. This allows me to pinpoint the offending hardware once I have found the pole.

When the utility crew came out, I observed the repairs being made–most while the line was hot. This, of course, saved shutting down the line. On one occasion the line foreman was standing next to me and listening to the ultrasonic noise coming from the Model 250 parabolic locator. At the exact moment the crew lifted the hot line off the offending insulator the noise stopped. The correction was dramatic and convincing.

STORY 15

THE CASE OF THE DUELING POWER COMPANIES

by Jim Hope

It was a hot August day when I received a call from a power company that had received a Public Utility Commission complaint, and also a complaint from the FCC on a customer in western Oklahoma. Two power companies were involved and had sent representatives to check out the problem, and both had made repairs to their lines over a period of six months. Both companies said that the other company was responsible for the TVI. The customer had become very tired of the finger pointing and had called the FCC and the Public Utility Commission to try to get his problem resolved.

I agreed to go out to find the problem and received full cooperation from both the private power company that called and the Co-op. When I arrived I met with the customer. He was a business owner who had a business in a small town served by the power company and a residence out of the town, served by the Co-op. We went to his residence where I observed severe electrical interference on his television. Since the customer lived almost seventy-five miles from the nearest television transmitter, the signal strength was not the best, but it was hard to argue with the high level of interference that was being generated by the power line.

I proceeded out to my vehicle and set up my monitoring equipment. I was amazed at the level of interference that was present. The other thing that was unusual was the direction it was coming from. Both power lines that were in question were to the south of the customer's residence, and the direction of the interference was to the northeast. In the process of tracking this down I ended up 1/2 mile to the northeast at an oil well pumping site that was served by the Co-op. Upon further investigation I found a set of dead-end bells located on transformer pole that were literally melting and dripping on the ground.

I then returned to the customer's residence and informed him as to the location of the problem. He asked me to talk to the people at the

Co-op to see if I could get them to fix the problem. I agreed and he made the call to the Co-op. They sent a representative out to meet me, and he began telling me that he had patrolled all of the lines in the area, and that the only noise in the area was on the other power company's line across the street to the south. When I asked him what equipment he used to determine this, he told me that he always used the AM radio that was in his company truck.

I then explained to him that his customer was complaining of television interference, not AM radio interference, therefore he was using the wrong test equipment. He agreed to send a crew to solve the dead-end bell problem but was not sure that it was going to solve the customer's problem.

Upon changing the dead-end bells the complaining customer had no more problems, (and the Co-op was saved an outage).

The representative from the Co-op asked what equipment I used to locate this problem. I explained that I used a multi-band field strength meter and a directional Yagi antenna. This is what Radar Engineers call their Model 237 DDF locator. I then used an ultrasonic detector with a parabolic reflector to locate the specific problem on the pole. This was made by Hewlett Packard and is no longer available. [A much more sensitive instrument ultrasonic instrument designed specifically for locating power-line sparking is now available, the Model 250 from Radar Engineers. ed.]

When I left western Oklahoma, I think that the customer was happy, the Co-op was happy, the power company was happy, the P.U.C. was happy, and the F.C.C. was happy!

The moral of this story is that it takes the proper equipment as well as the proper knowledge to solve each problem. There is no one piece of equipment that can find all cases of interference. I carry no less than a dozen different pieces of equipment in my vehicle and have used all of them on different cases of interference around the United States.

[Note 1. In this case, Jim knew where the problem was before checking with the ultrasonic locator from ground level. However, considering the power companies were dueling, he wanted to be able to say he had located the structure at UHF frequencies and pinpointed the source on the structure with ultrasonic frequencies. *Usually* the location would be confirmed *on* the structure by a lineman using the RF-ultrasonic hotstick sniffer.]

STORY 16

LIGHTNING ARRESTER IS NOW IN THE MUSEUM

by Mike Martin

I received a complaint from a PEPCO customer who was also an amateur radio operator. He was complaining of RI. The unique thing about it was the problem occurred in the rain as well as in dry weather. I had him tune in the noise while I was on the phone, and sure enough, there it was--"power-line noise." If I've heard one, I've heard a million power-line noises (even if they do not all sound alike). However, there was still the factor of rain, which it wasn't doing at this time.

I told the customer he may very well have two sources, because the likelihood of power-line noise in the rain is somewhere between slim and none. [Watch out for Murphy's Law!]

I asked the customer to call the next time it rained, which was supposed to be the following day. Well, what do you know, it <u>did</u> rain, and he <u>did</u> call, and as it turned out, it <u>was</u> power-line noise. I made an appointment and arrived with an O.H. (overhead) crew. I located the pole the source was on with the help of my DDF equipment. I then put the O.H. crew to work with my 247 hot-stick line sniffer. Using the VHF rod of the 247, O.H. was able to narrow down the area of arcing to the lightning arrester and the cutout on the pole. However, they weren't able to pinpoint the exact source using the <u>ultrasonic</u> probe.

Considering the factors involved, wet weather and not being able to detect it with the <u>ultrasonic</u> probe (which is direct line-of-sight), I determined the source of arcing must be inside a device. The cutout is all external except for the fuse link. However, the hollow tube around the fuse link is open at the bottom, and the ultrasonic usually detects noise there, if present.

With no detection at the fuse link, I decided to go with the lightning arrester. O.H. disconnected the primary tap to the arrester and the noise stopped. We reconnected it and it returned. The arrester (See Fig. 9-13) was replaced and the AM radio noise cleared up.

That lightning arrester had made its way into the Museum of Power-Line Noise Sources.

Fig. 9-13. Lightning Arrester Containing Noise Source.

Laboratory tests made on the above lightning arrester indicated that when the ususal operating voltage (7.2 kV) was applied, the arrester was quiet until it was tapped lightly. Then it would intermittently radiate noise which could be detected at 500 MHz. Internal sparking was obviously taking place.

The above lightning arrester is also shown on page 5-2, Fig. 5-10. A discussion on locating lightning arresters is given in Chapter 7.

STORY 17

UNDERGROUND NOISE REVEALED OVERHEAD

by Marv Loftness

A power company called me to help locate the cause of a high-level radio noise "covering" a small town in western Washington. This is in a mountainous area where radio reception is none-too-good, but important, if for no other reason than for weather and school bus reports.

The company suspected an overhead corona problem resulting from the recent upgrading of a 12.5-kV line to 34.5-kV (this proved not to be the cause). They said AM radio noise was all over a mile or more of line, and they were having trouble locating the source. Fig. 9-14 shows a section of this line containing the source that was located.

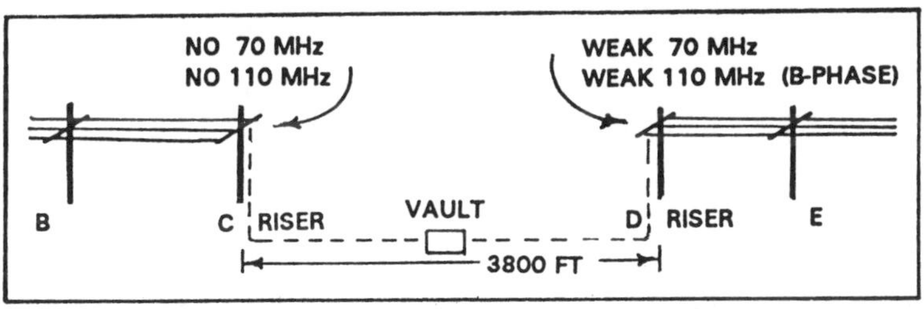

Fig. 9-14. 35-kV Line Containing Noise Source.

On this job I was accompanied by a power company technician. On our first pass along the suspected section of line, using the scope and vehicle method, *with VHF observations,* we observed no interference sources present on the overhead lines. This proved to be correct: there were *no sources* on the overhead -- even though there was a high level of AM radio noise *reradiating* all along the line.

This line was partly in a wooded, brushy area, and on the first pass we did not get as close to all of the line as we wanted. However, on the second pass, with special effort, we were able to get close to the two riser poles in a section of the line. Here we would look for noise at the highest possible frequencies. At one pole (D) and one pole only, we noted a weak noise at 70 MHz on our scope.

With line crew help, both riser poles D and C were sniffed out with the 247 hot-stick sniffer, and pole D only indicated output at 110 MHz; this was a fairly weak output, as you would expect from a site of reradiation, rather than from a site of radiation, at a source. (This reradiation we could detect on B-phase only.)

The underground cable fed by this riser was then the likely suspect. We proceeded to the underground vault 1900 ft away (shown in the figure) and, after opening the vault, the maintenance crew found with the sniffer, at a B-phase elbow-disconnect, high noise radiation at 110 MHz.

With a hot-stick tool, a crewman firmly pushed in the disconnect so that it was properly seated, and the radio noise over the wide area suddenly disappeared! This, of course, generated a great sigh of relief among the people of the village AND the power company.

CAUTION: Equipment in an undergound vault or pad-mount looks very innocent. Do not get close. Voltages are dangerously high. Leave sniffing and other work here to certified electrical workers.

STORY 18

GOES WILD WITH NOISE

by Marv Loftness

The power-line RFI investigator frequently has the opportunity of locating interference where the source is located in a home not far from away from the complainant's home.

In this example, I was called to help a power company with a "difficult" intermittent TVI complaint. Arriving on the scene I found that yes, it was an intermittent problem. With my vehicle parked across the street from the complainant's home (the Kirby's), I noted bursts of noise that varied in their length and spacing. After the first couple of bursts, I got a direction on the source by observing the noise pattern on my oscilloscope and by quickly rotating my antenna to peak on it. Because of the way the noise was acting, and the way it looked on the scope, I knew it was electrical and suspected it might be a doorbell transformer. I then proceeded to a location a couple of houses south of the Kirby's, as directed by my scope, and waited for another burst, which led me to turn my antenna toward the Smyth residence.

Suspecting something in the Smyth's house, I walked up to the house *and pressed the doorbell button. The sniffer went wild with noise* the minute I pressed the button. Mr. Smyth came to the door, and in response to my question, said, "Yes, we do have interference and on all channels. It just came on again and is usually intermittent."

The fact that Mr. Smyth had interference on all channels, of course, meant that I was quite close to the source.

I then advised Mr. Smyth that I thought the noise source was in his house, and that it was probably his doorbell transformer. He was somewhat surprised at this, and then told me that he thought the transformer was in the attic, and it would be hard to get to. The noise then went off again, and I went to my car to get my hot-stick sniffer, in order that I might scan the ceiling with it.

I returned with the sniffer and (with the noise on again) extended its antenna rod and scanned the ceiling until I located a noisy area. This was an area about 3 ft in diameter, and I advised Mr. Smyth that he or his electrician would find the noisy transformer in the attic above that area.

We then turned off the service-entrance breaker connecting the doorbell transformer, and all was quiet and stayed quiet. Mr. Smyth then said he would leave the breaker switch off until the transformer has been replaced. He was appreciative and happy to have good television restored.

I then went over to Mr. Kirby's, who said the interference was <u>bad</u> intermittently but now was now OK.

After waiting about 20 minutes at the Kirby's with no interference observed, it was apparent that this particular interference problem had been found, good TV reception had been restored, and the case resolved. Most interference investigators have found more than a few defective and noisy doorbell transformers, and would not be surprised at this story.

A typical doorbell circuit and its transformer is shown in Fig. 5-21, Chapter 5.

STORY 19

THE UNEXPECTED "DROP-OFF"

by Marv Loftness

The noisy structure shown in Fig. 9-15 was quite easy to locate. This was in the investigation of a TVI complaint. After leaving the customer's residence, it took less than ten minutes, going in the right

Fig. 9-15. RI/TVI Source Structure, Partial View.

direction, and guided by my scope, to identify the source pole, and to identify a noisy area on the pole. I also expected it would be easy to pinpoint the source with the hotstick sniffer. However, before the line crew arrived, they had sectionalized the line to de-energize the structure, so there would be no way to pinpoint it by RF or ultrasonic means. However, we did pinpoint it, as you shall see, by more dramatic means.

The source location had been narrowed down to being in the **A-B** area shown, with an 800-MHz UHF observation from the ground.

I found this to be a steady source and, I should mention, I did not shake the structure with its guy wire, as I sometimes do to activate an intermittent source . **Caution:** A guy wire is sometimes safe to touch and sometimes not--shaking a structure with its guy wire should be left to power company people!

When the crew arrived, after they had de-energized this line section, I asked them to tighten up everything, especially in the (**A-B**) area. On a job such as this, one of the linemen might say, "I can tell you right now, it is one of those pin insulators," pointing to the <u>**B**</u> area. This he bases on a previous experience. For some reason, people sometimes get "hung up" on a certain condition as being "the" source. Some might say it's a loose tie wire; some, a loose crossarm brace; and some, a loose hot-line clamp; this might be right but often wrong. Then the lineman in his service bucket went up to start tightening hardware in the suspected area, one of the first things he did was to take a hotstick tool and attempt to tighten the hot-line clamp at <u>**A.**</u> He barely touched the clamp with the tool, when the clamp and its jumper *fell off the conductor!*

[As you can imagine, I was <u>very</u> glad I had not shaken the structure while it was energized. This could have been embarrassing, to say the least].

A little frantic activity ensued because it was found that the approximate 75-ft span of the ASCR (aluminum-conductor steel-reinforced) primary conductor cable was suspended by <u>one strand</u>

only! The line crew quickly applied a come-along on the line to span the clamp area and, to support the conductor. They then proceeded to splice-in a section of new conductor.

After some general tightening of hardware, the line was re-energized and the structure was found to be noise-free, and the customer's TV was free of sparking noise.

I requested that the line crew donate me the old section of conductor and the old hot-line clamp, which they did--so I that could take the pictures of Fig. 9-16A and 9-16B.

(A)

(B)

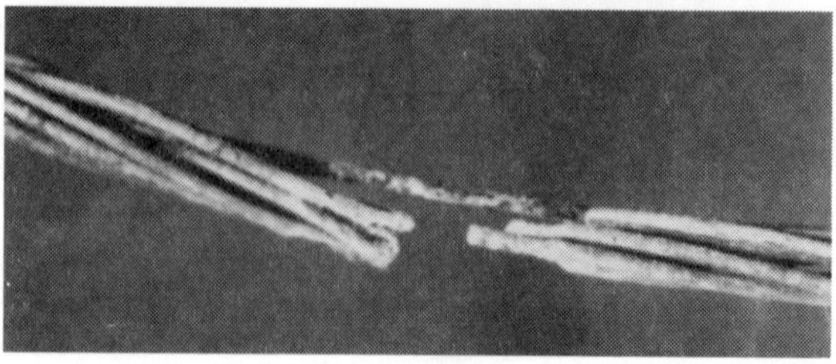

Fig. 9-16. (**A**) Loose Line Clamp Causing TVI. (**B**) Single Strand supporting Conductor in Road Crossing.

A hazardous situation!

The old hot-line clamp is shown with the outer aluminum strands corroded through to the steel core of the ACSR (aluminum-conductor steel-reinforced). Fortunately, this steel strand was good quality steel.

This story, of course, illustrates a poor way to pinpoint sources. However, the location and correction did result in a satisfied customer (...but in some shook-up linemen and an RFI investigator?)

STORY 20

A WEIGHTY NOISE PROBLEM

by Marv Loftness

As mentioned previously, various people at various times get "hung up" on where noise sources are most likely to be found on power lines, and also on what is the best way to correct them. I was no exception, and in one of my early locating excursions for the BPA, I was working on a 69-kV line which had suspension insulators. At the time I was hung up on the idea that hanging weights on insulator strings would be the answer to all my noise problems. Fig. 9-16 shows a typical structure on the line in question, with a 100-lb lead-ball weight hanging on each insulator string.

Fig. 9-16. Weights on Suspension String, 69-kV Structure.

Anyway, this was a very noisy line, and I had a very cooperative line crew who hung weights on about 14 poles of line, generally 3 on each structure.

Of course, as you have guessed, after the weights were hung, we still had the noise (at least most of it).

Then on a nice, sunny day, with my jeep windows open, I decided to drive along the line, stop at each pole, and tap it with a sledge hammer to see if I could "interrupt" the noise to identify a source pole. The line was in fairly open fields, with a maintenance trail, but called for a 4-wheel drive in some areas.

Near the end of the "treated" portion of the line, I was getting a little tired of the exercise and decided that instead of getting out to hit the poles, I could gently tap them with my jeep--a brilliant idea.

At the first pole where I tried this new method it worked, but in an unexpected manner.

I lined up my jeep with the pole, approached it and gently "tapped" it while watching my scope to see if the noise pattern would be interrupted; it did flicker, but I also heard in my left ear a distinct whooosh! Then somewhat of a *thud!* Strange, I thought. Opening the door, I looked down and saw a hole in the ground about six inches deep; I looked up and saw that a lead weight was missing. I poked into the hole, and sure enough, I struck lead. I dug it up and put it in back of the jeep for return to the line crew. I don't remember that it weighed anything, as I was partly traumatized.

At the time I did have my hard hat on, and the roof of the jeep was metal, but if the weight had gone through the roof it might have hurt a little bit.

This would not have happened if I had by that time invented my hotstick line sniffer. In that case, the line crew would have sniffed out the poles and found several actual sources, which I suspect now were also occurring at several hardware locations, besides at insulator connections.

How did the story end? I was embarrassed, of course, but oddly, not criticized--rather commended, believe it or not, by the safety section (who thought it was a good story for their safety sessions).

As you have noted throughout this book, there are many possible locations, as was shown in Table 5-2, Chapter 5. The first 12 locations listed in this Table are the <u>most likely</u> sources, but it is difficult to list these in order of occurrence because the order differs with the practices of various companies, particularly with hardware used, devices used, and environmental conditions.

The method of hanging the lead weight is shown in Fig. 5-5, Chapter 5.

Chapter Ten

FOUL WEATHER POWER-LINE INTERFERENCE

Much of this chapter is material excerpted and updated from the author's handbook, "A Practical Handbook for the Location, Prevention, and Correction of Television Interference from Overhead Power Lines," written for the Bonneville Power Administration. (This book is now out of print).

RFI on Transmission and Distribution Lines

On transmission lines, corona-related RFI is an important consideration. This has been especially true since the advent of extra high voltage (EHV) lines--those lines operating at 345 kV and above.

On distribution lines (and on transmission lines operating at 230 kV and below) "*sparking*" accounts for nearly all power line interference.

The presence of *corona* on EHV lines often indicates the presence of RI, but no TVI. Under certain circumstances, however, serious TVI can also result from such corona.

EHV Corona

Although corona was discussed in Chapter 2, a further explanation is given here as an introduction to the special study of EHV-corona problems.

When air breaks down as a result of high-voltage stress in the air immediately surrounding an EHV conductor, electrons are stripped from the nearby molecules in this stressed air. This results in free electrons and "charged" ions of the atmospheric gases. An electrical discharge from the power-line conductor into the surrounding air then takes place, with light emission as well as audio and radio-frequency noise. This discharge we call corona.

During moderate rain (with thoroughly wet conductors), and from a vantage point where a long transmission line span can be seen, the entire span may at first appear to be aglow.

Upon close examination it will then be seen that thousands of corona sources are present along the conductors. These will be present in two forms, impingement pluming and spray pluming.

The impingement plumes result from electrical discharges between raindrops and the conductor as these raindrops approach the conductor. These will be seen as flashes on the top and sides of the conductor, as shown in Fig. 10-1.

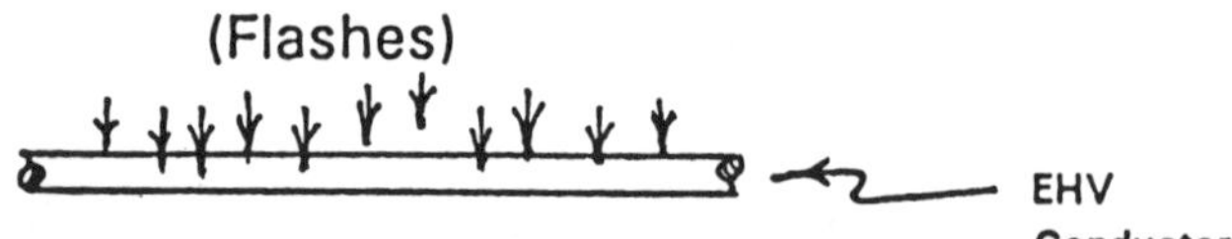

FIG. 10-1

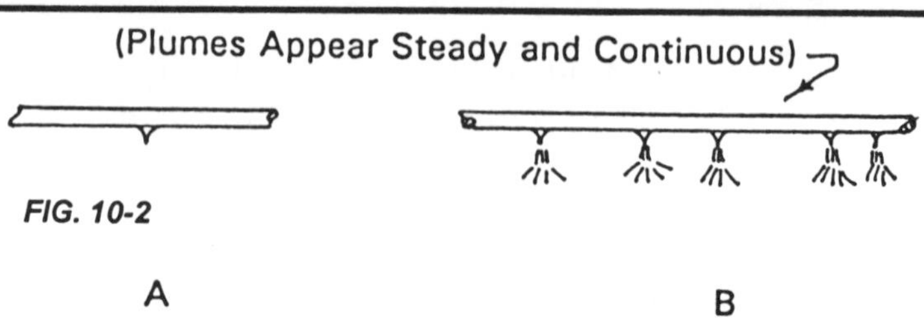

FIG. 10-2

In Fig. 10-2A, a single water droplet is seen forming on the bottom side of an EHV conductor. Such water droplets accumulating on the conductor are distorted into pointed shapes by action of the electrostatic field and provide protrusions for corona plume formation, as shown in B of the figure (10-2).

Transmission line corona can be observed visually with the aid of binoculars. However, the light seen from corona, which is near the ultraviolet end of the spectrum, is very weak and almost complete darkness is required. After one's eyes have adapted, many interesting observations can be made on a EHV lines.

The spray plumes forming on the underside will be more obvious than the impingement. These comprise the main corona activity responsible for the high level of RI and TVI (as well as audible noise, AN) on certain EHV lines in foul weather.

The droplet spray plumes will appear steady and continuous to the eye, in contrast to the flashes of impingement plumes. An individual droplet plume may last for several minutes and appear continuous during this time--it may migrate, however, moving slowly down the sag of the conductor span.

When rain first commences, RI will be observed resulting from the impingement plumes, but after water droplets accumulate on the bottom side of the conductor, this noise will be overridden by RI from the droplet spray pluming.

In the absence of wind, water droplets may hang for hours on the underside of *non-energized* conductors; however, on energized conductors they are dissipated by corona action and conductor heating in a matter of minutes. To illustrate, twenty minutes after cessation of rain, the RI noise level from a 500-kV line was found to be reduced at least four times (12 dB) as a result of the reduced number and intensity of plumes.

It will be observed that the amount of RI generated by a plume is relative to its length, with a long plume generating more noise than a shorter one. The noise radiated by a particular span of conductor then depends on the plume lengths as well as their number. After a certain number of plumes are established on the conductor, increasing their number has less and less effect upon increasing the overall noise level.

Most of the author's EHV observations reported in this chapter and in Chapter 11, were along a line with horizontal spacing of the phase conductors as shown in Fig. 10-3. Here the line conductor is a 2½ inch diameter expanded ACSR.

Fig. 10-3. 500-kV EHV Line with Horizontal Phase Spacing.

On transmission lines the possibility of a corona discharge taking place at the surface of a conductor is dependent upon several factors. As these factors change the noise level will change or come and go.

1. Power line operating voltage
2. Conductor surface voltage gradient.
3. Conductor diameter.
4. Conductor surface conditions
5. Atmospheric conditions
6. Weather conditions.

Fig. 10-4 shows an oscilloscope photo of EHV noise at 75 MHz. The photo was taken near an outside phase (1); this phase shows the highest noise level, and phases (2) and (3) lower levels.

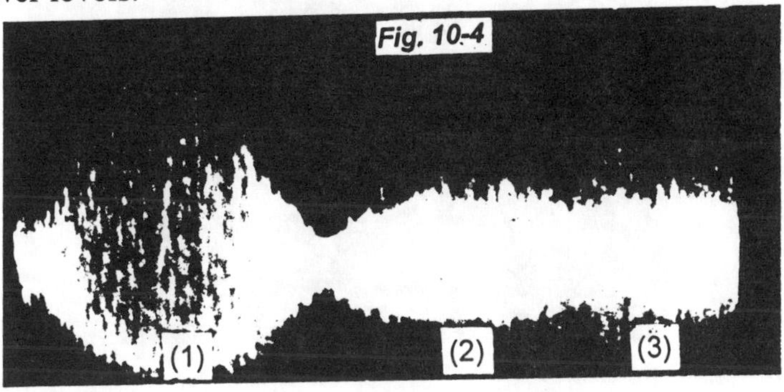

Fig. 10-4. Corona Noise seen on Oscilloscope Near 500 kV EHV line.

The typical three bands of "foul weather" TVI resulting from the 3-phase conductor of the EHV line are clearly seen on the TV in Fig. 10-5. Here a "raster" only is present on the TV screen; this raster includes synchronizing information transmitted by the TV station, but no picture. The TVI with a picture being received is shown in Fig. 10-6.

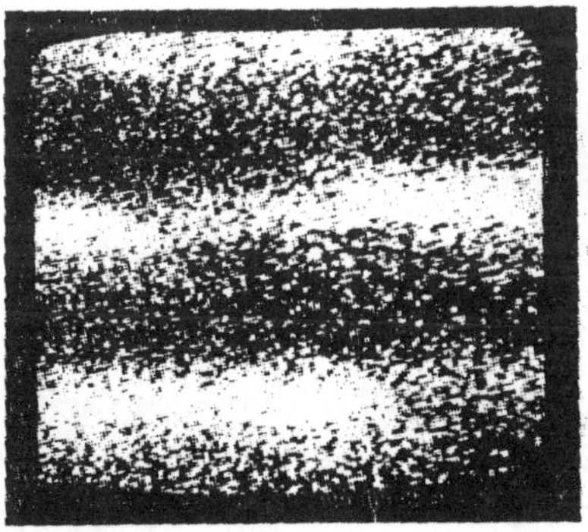

Fig. 10-5.

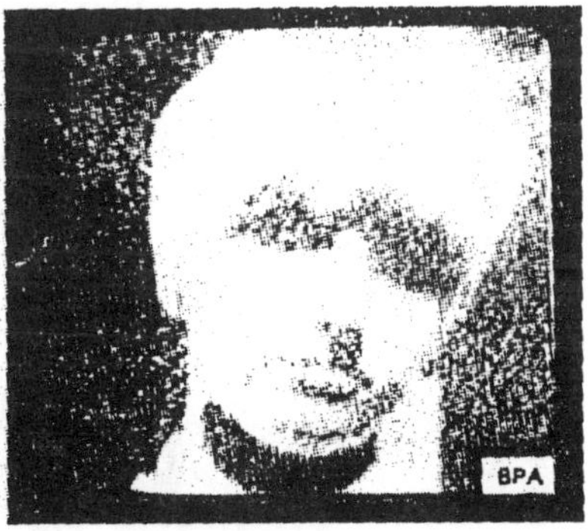

Fig. 10-6.

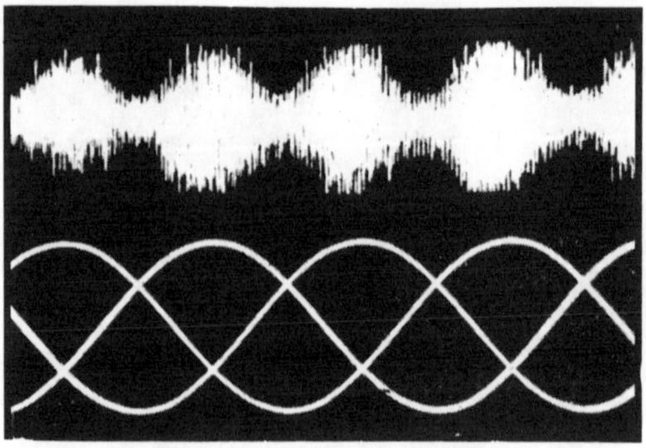

Fig. 10-7. Drawing Showing EHV-Corona Noise– as it might be seen on an oscilloscope. This is on Positive Alternations in Foul Weather, on a Three-Phase Line.

Surface Voltage Gradient

The voltage gradient refers to the voltage difference between two points divided by the distance between them. A, B, C & D are equipotential lines and in Fig. 10-8 the highest gradient exists in space A , the lowest exists in D.

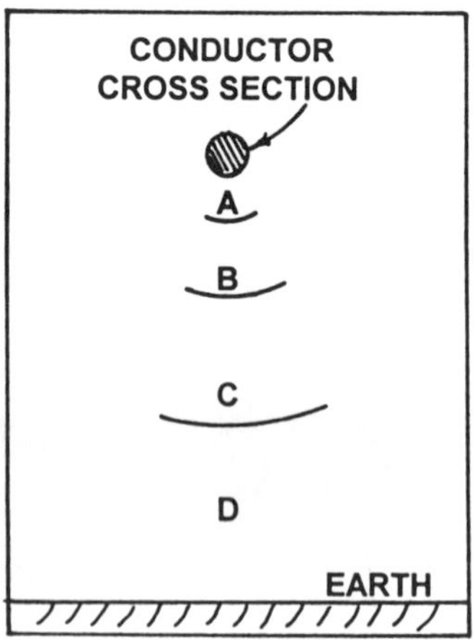

Fig. 10-8. Voltage Gradient.

For 500-kV lines the voltage gradient at the conductor surface may exceed 15 kV per cm, that is, 15 kV per the first centimeter out from the conductor surface, and we call this the surface voltage gradient. With a surface voltage gradient of 15 kV per cm, the insulating property or "dielectric strength" of the air near the conductor is obviously highly stressed. The surface gradient line is not shown in the figure.

Voltage Gradient Considerations

In addition to conductor diameter, the conductor surface-voltage gradient depends on system voltage, the presence of an overhead ground wire, spacing of the conductor to earth and to other phase conductors, and the number of conductors per phase (in a bundle). When the parameters of a particular line design are proposed, computer programs designed by Vern Chartier[10] while he was with BPA, are normally used to calculate the surface gradient for each of the phase conductors resulting from proposed parameters.

Conductor Diameter.

A larger-diameter conductor will have a lower voltage gradient at the conductor surface. Another important, but not necessarily desirable, feature of the larger conductor is that *the voltage gradient outward from a large diameter conductor decreases less rapidly than it does from a small diameter conductor.*

Conductor Surface Conditions and Corona Formation

The following factors have a strong influence on the generation of corona: the type of conductor, that is, normal stranded ACSR (aluminum conductor, steel-reinforced) or ASCR/COMP "compacted-round" (or other); the condition of its surface, that is, the degree of smoothness or roughness; the presence or absence of pollution and the presence of surface grease or water droplets.

A transmission-line conductor when first strung will usually have high corona activity due to surface irregularities such as aluminum burrs, surface grease, and splice compound. After about three years and the conductor is blackened, the RI level in fair weather is reduced and the foul weather TVI level reduces to minimum. Even after the first year the corona level is noticeably reduced. As the conductor ages, it more readily draws up water into its strands. With new conductors, water beads form more readily on the surface.

Atmospheric Conditions

A reduction in the barometric pressure or an increase in the ambient temperature can reduce the air density. This in turn reduces the breakdown strength of air and increases the likelihood of a corona discharge taking place on a conductor. This can be important at elevations of approximately 7000 ft and higher. There are a number of transmission lines at these elevations in the United States and Canada, but these have relatively few residences near such lines.

Weather Conditions

A drastic change in corona generation and consequent radiation from a power line can occur with a change in the weather. This noise radiation consists of radio, television and audio noise. It leads a nearby resident to identify the power line as the source of his TVI. When he hears the "same noise" come up on his radio as the audible noise he hears with no radio, he properly blames the power line for both his radio and television interference. Further, one complainant even reported that he could *see* the three power lines on his television screen. Referring back to Fig. 10-5 it appears that this may not have been be too far from the truth!

Extent of EHV Foul Weather TVI

The presence of corona on EHV lines, however, often indicates the presence of RI but no TVI. Under certain circumstances, however, serious TVI can also result from such corona.

Foul weather TVI is usually limited to residences within about 600 ft of the transmission lines. In outer fringe reception areas such interference may be noticed out to 1500 ft or more. In

Grade A reception areas, the TV signal strength is sufficient to override the TVI even at the edge of the right-of-way.

Foul weather TVI occurs in rain, fog or snow. Wet snow generates the highest TVI level. Considering complaints received, this foul weather TVI is only a problem on EHV lines. With sophisticated test equipment the corona radiations at TV frequencies *can* be observed at low levels directly under a 115-kV line, but at this voltage it is not a cause for concern.

> For the observations made in this chapter, a standard frequency of 75 MHz was used. The patterns at TV channels 2, 3, 4, 5 and 6 would be of the same shape.

Although the vertical scan rate for television pictures is a *nominal* 60 times per second, it is actually 59.94 times per second. This difference from the power-line frequency of 60 Hz accounts for the upward drift of the interference, the "snow" patterns caused by the foul weather corona radiations. This *drift* we also noted in Chapter 1 as occurring with sparking RFI.

The sound portion of the TV transmission is usually not affected by power-line interference. The sound is frequency-modulated and is immune to the *comparatively* "low level" foul weather radiation.

The picture (video) information is amplitude modulated, occupies a much larger bandwidth, and is highly susceptible to low-level interference. "Low level" here means low absolute level: for comparisons, in a Grade A signal area (in or near the TV station transmitter) a received TV signal strength of 20,000 microvolts per meter (μv/m) may be found; in a fringe area, a signal level of 100 μv/m might be a typical value, and here a 10 μv/m of corona noise will cause serious interference. 100

microvolts per meter of signal will provide good TV picture, in the absence of noise.

Because of the huge gap in assigned TV frequencies between channels 6 and 7, we see no *corona* interference in foul or fair weather at channel 7 (174 MHz) or higher channels.

Bundle Conductors

Bundled conductor designs can be used to reduce the corona and the interference levels from EHV lines, particularly in foul weather. These emissions include radio interference, TV interference, and audio noise (AN).

The use of bundled conductors increases the effective diameter of a "phase conductor," (and we might call it a “bundled phase”), thereby reducing the surface voltage gradient. Also, as previously pointed out, the rate of gradient decay outward from an individual small conductor is rapid; and very important, when a small-diameter conductor is in a bundle, the rate of gradient is still rapid for the small conductor as well as for the entire bundle. This combination of lowered surface gradient and increased rate of gradient decay reduces considerably the incidence and size of corona plumes formed on the conductor, and thereby considerably reduces the corona currents and the resulting RFI level. A typical bundled-conductor phase is shown in Fig. 10-9. Here the 1.3 inch diameter conductors are held approximately 18 inches apart by the aluminum spacer.

Shown is a 3-conductor bundle, which in the author’s opinion is the most suitable for 500-kV EHV lines. A two-conductor bundle may also be used, but it is subject to aeolian vibration and generating weird audio noise. A 4-conductor bundle is also sometimes used, and its only advantage may be that of being able to handle more current

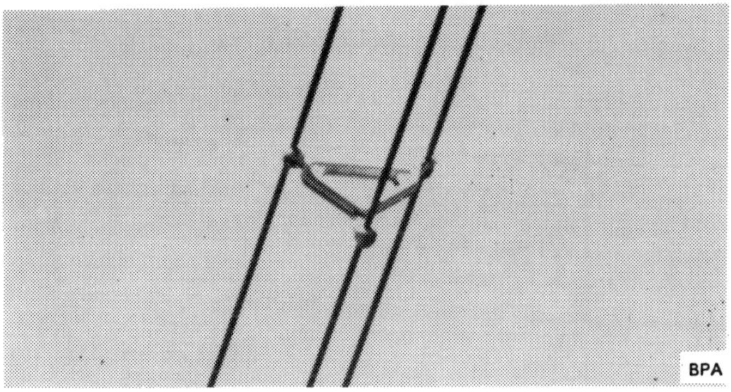

Fig. 10-9. 3-Conductor Bundle--A Typical 500-kV Configuration, with 18-inch Spacing between Conductors.

Fair Weather Corona

You may, on occasion, get a complaint of EHV *audio* noise in *fair weather*, as well as RI complaints of radio amateur or AM-radio noise. The AN is likely from corona plumes on pointed objects on the line, such as frayed conductors, hardware burrs, or "trash on the conductor." The RI here may be *from either sparking or corona.* If it is sparking, of course, it will be detected up into the UHF range, whereas the corona noise will not.

Heard with the unaided ear, the plumes may be sound quite noisy–with long plumes having a loud crackle. A group of plumes may cause a roar.

You will not see these plumes in the daylight, of course, but individual plumes can be pinpointed with an ultrasonic locator such as the RE Model 250, which was shown in Chapter 7, Fig.7-1. Such a device detects the 40-kHz component of the corona radiation and converts it to an audio frequency.

Fair weather versus foul weather

EHV-noise observations are shown in Fig. 10-11.

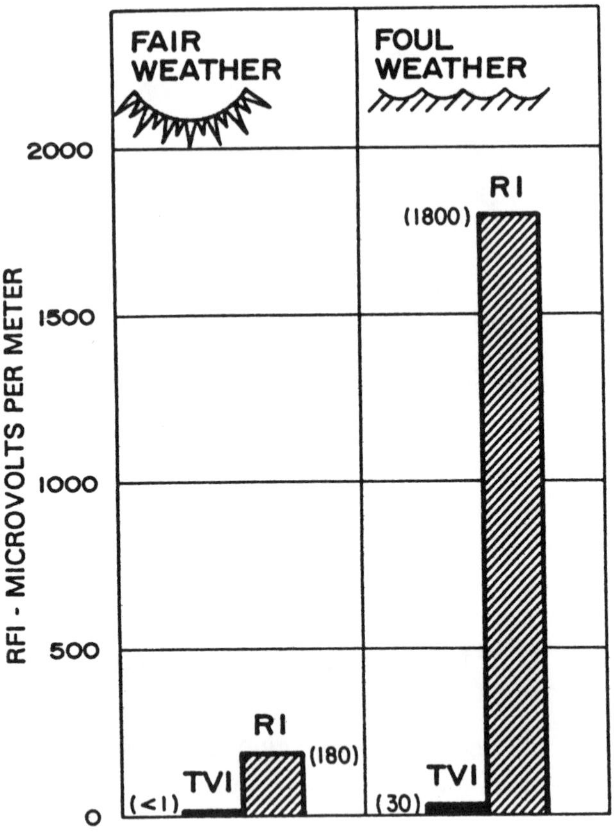

Fig. 10-11. EHV-Line Noise Comparisons, TVI at 75 MHz, vs RI at 1 MHz, in Fair Weather vs Foul. This is approximately 200 ft from outer phase of a non-bundle 500-kV Line as was shown in Fig. 10-3. Typical values are shown. From Loftness-Clark [11].

A 525-kV Line of optimum design from the standpoint of appearance, RI, TVI, and AN is shown in Fig. 10-12.

Figure 10-12. Spacers on the 3-Conductor-Bundled Phases of a 525-kV Line–here shown highlighted by the late afternoon sun. This is a Bonneville Power Administration line near Salem, Oregon. [CHARTIER-LOFTNESS OBSERVATION SITE]

SUMMARY–CHAPTER 10

- In foul weather, corona on lines operating at 345 kV or above may cause RI and TVI.
- Water droplets hanging on a conductor readily support corona pluming.
- Thousands of plumes may be observed on a single span of 500 kV transmission lines in foul weather.
- Corona is a "partial" electrical discharge into the air from a high-voltage conductor.
- Corona emission from a conductor is dependent upon: surface voltage gradient; conductor surface conditions; diameter of the conductor; moisture; rain; snow; and atmospheric pressure.
- Foul weather TVI is usually limited to less than 600 ft from the EHV line.
- On a TV, foul weather TVI appears as three horizontal noise bands drifting up across the screen.
- Bundled conductors can be used to reduce corona emission.
- In fair weather, *EHV lines will not cause TVI,* but may cause RI at some *radio amateur frequencies.*

Chapter Eleven

TVI CORRECTION NEAR EHV LINES

Introduction

Television reception can be improved at almost any location by increasing the signal-to-noise ratio (S/N ratio). Near EHV (Extra High Voltage) power lines this often involves special problems, and certain methods for improving the S/N ratio which have proved successful will be discussed. This chapter will describe the correction taken to improve this ratio at the receiving TV set to improve this ratio, particularly when it occurs in foul weather. This discussion is excerpted and updated from the author's handbook on the subject written for the BPA [5].

This chapter is of interest not only to power companies. It also contains very good information for cable TV companies when locating sites suitable for good VHF TV reception.

For the home owner, recommendations are given on how TV reception may be markedly improved with antenna and system refinements, for off-the-air TV reception.

For the radio amateur, it shows why he may be receiving VHF noise from EHV power lines during foul weather.

Nowadays, of course, in very many cases, the power-line TVI correction can be made with satellite reception. However, a surprising number of people *do rely* on off-the-air reception, and with them, local channel reception is a must.

Television Interference Correction

Power lines or structures can cause at least four types of TV interference: fair weather sparking, foul weather corona, signal reflections, and signal blocking.

Fair weather sparking can be corrected on the line once the sources are located. *Foul weather corona television interference* cannot be economically corrected at the power line source. It can be corrected by lowering the operating voltage or reduced by reconductoring the line (at great expense), but these alternatives are usually not acceptable to the power company. *Ghosting* and *blocking* are of relatively minor concern but do occasionally occur .

In this chapter it will be shown how television reception can be improved where there is little prospect of removing the interfering sources, as in the case of foul weather TVI, ghosting, and blocking.

Fig. 11-1. Example of Residences in a Low Signal Area Requiring Correction for Foul Weather TVI.

Good television was restored at the residences of Fig. 11-1, with a remote antenna located approximately 400 ft from the power line, and serving the homes with a small cable TV system.

Fig. 11-2 shows an example of an EHV corner structure which was blamed for causing reflections; here correction at a nearby residence was accomplished with the installation of an antenna 50 ft from the residence.

Fig. 11-2. 500-kV Corner Structure Blamed for Reflections at Nearby Home.

Signal-to-Noise Ratio Requirements

The question is often asked as to what signal-to-noise ratio (S/N ratio) is required for good television. A ratio of 100:1 is a good answer.. In the case of foul weather interference it will be found that a 100:1 ratio *will* provide good reception. To determine the ratio, a Sadelco-type FS4 or FS5 field strength meter was used. These instruments have bandwidths of approximately 0.5 MHz and were selected because they have normally been available to television cable company technicians and antenna installers.

With your test antenna located at your proposed antenna site and pointed towards the TV station, you should receive 100 times (40 dB) as much signal as you do noise.

At the l00:1 ratio, using one of the above meters, foul weather TVI will be barely noticeable, and from the author's observation, such a ratio will not cause the viewer to initiate a complaint.

Critical Zone Near Transmission Lines

A concern expressed is, "How far will foul-weather noise radiate?" And more important, how far must a home be from the transmission line in order *not* to experience foul weather television interference. This is just as difficult to answer as the S/N question, and the immediate answer is, "It depends." It depends on a number of factors which will be discussed.

Transmission lines, where possible, are routed through areas of low population density. When some of the first lines were constructed for 500 kV, it was thought that residences near the edge of the right-of-way would possibly be affected, but surely none beyond 600 ft. Fortunately, only a moderate number of residences beyond this arbitrary 600 ft have been found to have valid EHV TVI complaints, as compared to those closer to the line.

The distance from a transmission line required to obtain a 100:1 S/N ratio depends upon:

(1) The intensity of the foul weather noise radiated from the conductor at the frequency of the TV channel being received.

(2) The television signal strength in the area.

(3) The direction the receiving antenna is pointing (that is, whether it is pointed into the conductor or away from it).

(4) The front-to-back ratio of the antenna (receiving TV *or* noise).

(5) The directivity of the antenna array.

(6) The efficiency of the antenna system, including the antenna, lead-in, splitter and other devices.

(7) The elevation of the conductor above average terrain and its elevation with reference to the receiving site (This will affect

the noise propagation between the conductor and the receiving antenna).

If some of these many factors are unfavorable, the only solution for correction may to be to install a remote antenna at a distance from the line where the signal to corona noise ratio is down to 100 to 1. In one unusual case, foul weather TVI on channel 2 was observed at over 2000 ft from the line. Here foul weather corona radiation was from a twin-bundle conductor 500-kV line. The signal strength on channel 2 at the residence was less than 75 microvolts, if measured on a dipole antenna.

The residence was in a valley with its antenna looking directly into a transmission line structure on a ridge of the valley.

Correction was obtained by locating horizontally stacked antennas at an optimum signal location about 150 ft from the residence, with the antenna array about 50 ft higher than the residence antenna.

Ghosting and Blocking

The validation of a ghosting or blocking complaint to determine whether or not it qualifies for correction is not as straight-forward as that of a foul weather complaint. The investigator usually does not know what reception quality was present before the power line was constructed. A signal survey on both sides of the line, after the fact and using both a field strength meter and a TV set, will usually indicate whether blocking is present as a result of the power line's presence. Such a survey, incidentally, will often indicate that the power line is blocking signal to the complainant's residence. A signal survey may not be conclusive, however, because the signal level may be down significantly on one side of the line because of terrain obstructing the TV signal propagation.

Power-line blocking might be suspected by a viewer when coincidentally with power-line construction, the TV signal level is down. Also, the complainant's viewing habits may have changed and his favorite channel suddenly has become a channel that was always somewhat ghosty but he was not previously concerned because he seldom watched that channel. Further, ghosting may be present because of new structures, or high-rise buildings, or a change in TV transmitter location or antenna changes. These possibilities suggest that often reports of blocking and ghosting are not power-line caused.

Fig. 11-3. Signal-Reflecting Structures, EHV Line and Water Tank.

Fig. 11-3 is an example of a structure completed about the same time as a nearby power line. This huge water-tank is more likely to cause ghosting than is the new power line.

Where ghosting is the complaint, the investigator should thoroughly check out the receiving antenna system before concluding that signals have deteriorated as the result of new power lines. Connecting a second TV set on the same lead-in, for example, without a proper splitter may reduce the lead-in's efficiency as a signal transmission line (because of impedance mismatch) and cause the lead-in itself to act as a receiving antenna. Where the rooftop antenna picks up signals mainly from one direction, this lead-in will pick up undesired signals from many directions. Loss of main signal occurs and ghosting may result. This will be more noticeable on some channels than others. Outlets in several rooms, may result in this antenna effect and ghosting may be noticed.

After a general signal survey of the complainant's location and surrounding area, and after considering the complainant's report and talking with neighbors (this all requiring perhaps two or three visits), the investigator will have a good feel for whether or not blocking or ghosting is actually present as a result of the power line.

Correction Criteria

The criteria for correction, when undertaken, should be to provide the complainant with pictures as good as or better than the pictures he received before the power line was constructed. These "before" pictures might have been quite snowy or quite ghosty, but they were "good" to him. It is amazing what people will watch and call good television in the far fringe areas.

Satisfactory TVI correction in far-fringe or near-fringe areas usually requires providing good reception on two to five channels. As a part of the general correction, available FM stations should be placed on the system. This can be done at slight extra cost and will also help to alleviate possible AM radio complaints.

Non-Power Line TVI; Poor Reception

Power lines are often blamed for TVI originating elsewhere. Symptoms appearing to indicate power-line fair-weather sparking, ghosting, or foul-weather TVI, can result from a defective television receiving system or from interference generated by devices in the home. Overloading (excessive signal) can also cause apparent TVI.

Two complaints were received by the author stating that the television picture became snowy during rain. The characteristic of foul weather TVI is, of course, the presence of wide bands of snow in the picture, and these complaints sounded legitimate. The residences were at moderate distances from the line and in good TV signal level areas; for these reasons the complaints were somewhat surprising. It was found that in both cases the 300-ohm twin-lead from the antenna to the television receiver was badly deteriorated. When this lead-in became wet in rain it passed very little signal–not enough to override the inherent television receiver noise--and the result was snow in the picture. Replacing the lead-in corrected the problem and the "foul weather TVI" disappeared.

The location of sparking devices in the home such as aquarium heaters and doorbell transformers (as discussed in Chapter 5) is usually not difficult or too time-consuming, once the investigator is at the residence; if this is done, and the problem corrected, other corrective action taken (with regard to power lines) will be more fully appreciated.

In visiting various TVI complainants, some very poor antenna system installations will be found; others will be quite good, but it is rare to find the installation that cannot be improved in some manner (for improved picture quality). This is a plus for public relations because *the completion of a corrective installation almost always results in the resident receiving better television than he had ever received.*

A Further Discussion of Ghosting.

If it can be assumed that the ghosting we see is a "one-bounce" reflection, it is easy to calculate the extra distance the signal would travel via a particular reflecting object to reach a residence antenna.

For example, for a reflected signal to arrive at an antenna one microsecond (μsec) late, it has traveled 984 ft (300 meters) further than the main signal. This is because electromagnetic waves (including light and RF waves) travel 984,000,000 ft per second, or 984 ft per μsec, then if it appears that we are dealing with a "one bounce" reflection, it is easy to calculate the extra distance the signal would have to travel via a particular reflecting object to reach the TV antenna. [The probable case in fringe areas, however, is to receive several reflected signals via different paths].

The first symptom of ghosting on a TV screen is a set of vertical bars on the screen. These *vertical* bars are ghost images of the *horizontal* blanking and synchronizing pulses, and are illustrated in Figs. 11-4 and 11-5.

The relation of the horizontal blanking and sync pulses to the picture information is shown in Fig. 11-4. The blanking and sync pulses are normally "off" the picture, but where reflection occurs, a ghost image becomes visible.

In Fig. 11-5 both the blanking and the sync pulses are seen. The sync has a higher percentage of modulation and thus a darker stripe (the narrow dark vertical stripe overlays the broad lighter stripe). Here the ghost pulses here are arriving about 25 μsec after the main pulses, and the man's tie "image" is 25 μsec to the right of the tie in center-screen. *The picture area* on the screen is the 54 microsecond interval indicated in Fig. 11-4 by the arrows in lower portion of the figure.

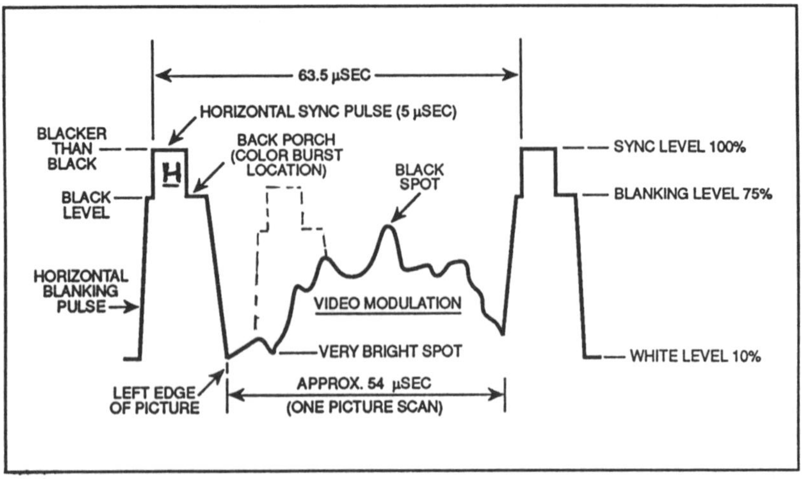

Fig. 11-4. Relation of Horizontal Sync and Blanking Pulses to TV Video Modulation. Ghost of blanking and sync pulse, normally not present, seen in dotted lines and (H) in Fig. 11-5.

Fig. 11-5. Ghost Image (H) of Horizontal Sync Pulse in Picture; Ghost Image (T) of Man's Tie. Note the picture is having some trouble synchronizing horizontally; this is indicated by the "wiggly" vertical lines.

Correction for Foul Weather TV

Correction for foul weather TVI resulting from the operation of nearby power lines can usually be made at reasonable cost by:

(1) Improving the antenna system at the residence.

(2) Installing a remote antenna (or antennas) at a sufficient distance from the power line so as to be unaffected by the corona radiations.

(3) Installing antennas for channels previously not received but acceptable to the complainant as alternates.

(4) Installing a satellite dish and system.

(5) Connecting a residence to an existing TV cable system.

Improved Antenna Installation

Site Location. The last place any compromise should be made is on antenna location. Every effort should be made to find and use the best possible antenna site (the site with the best possible picture reception) within reasonable distance of the residence. Several days spent in this search (if necessary) are well worthwhile because ghost-free and solid signals will save much effort in putting together a good system at reasonable cost. In most cases there is not much doubt about the best signal, and the search can be completed the first day. However, this location should be confirmed a week or two later in different weather to confirm the signal strength received.

In discussing antenna site location, the author is thinking in terms of rural areas where this is most often required (and where commercial cable TV is not available). Where problems in finding good signals occur, no possibility should be overlooked. For example, a building may be used as a shield, either from an unwanted reflection or from a noise from a particular direction. This is shown in Fig. 11-6. The antenna height is not the optimum for maximum signal strength, but the signal-to-noise ratio may be dramatically improved over that at a greater elevation; this ratio is much more important than absolute signal level.

Fig. 11-6. Channel 12 Yagi Antenna (A) Mounted So That EHV Noise Is Blocked by House–Single Channel Correction. Mounting mast is 1-1/4" diameter tubing TV mast (B).

[Note: No television set can provide good pictures without good signals coming in, and it appears reasonable that an expensive color television set merits an expensive antenna system if that's what it takes to provide a good signal.

If it is possible to get a hillside between the receiving antenna and the interference, this can be very effective. Shielding screens are effective, but other means for S/N ratio increase are usually adequate.

It may be of advantage to place the corrective antenna on a neighbor's property. This party may not have a valid foul weather complaint, but probably will be happy to permit (at no charge) an antenna installation on his property in exchange for connection to a system providing improved television reception. Fortunately, building a system capable of serving two or three houses often is only slightly more expensive than building one capable of serving only one residence.]

In searching for an antenna site it will sometimes be found that moving the test antenna 20 ft horizontally will make a significant difference in picture quality. This horizontal movement may be lateral to the TV signal path or along the signal path. Sometimes also, signals will come in fairly ghost-free at 10 ft above ground but be very ghosty at 30 ft above ground.

The best television signals are often found on a hilltop or on the far side (with respect to the TV station) of an open, fairly flat field. In a valley location, sometimes a notch in the hills will provide a more-or-less line-of-sight "view" of the TV transmitter, and lining up the antenna with this notch will provide a good picture. Actual line-of-sight for VHF television is not required as the signals do bend over the terrain more than might be expected.

Usually an increase in height at a particular site will result in stronger signals with no increase in ghosting, and after deciding on a site, it is worthwhile checking it out at the 50- or 60-ft height with a ladder or bucket truck for both signal strength and ghosting.

It is seldom necessary to place an antenna above the 50-ft height. Wood (treated) poles are recommended for antenna mounting.

It may be found that a desired remote antenna site is near a distribution line. If the option is available and loss of signal quality does not result, it is best to install the antenna on the side of the distribution line where it looks away from such a line. This will minimize the pickup of sparking radiations that might be present or later develop on this distribution line.

This sparking should, of course, be corrected, but depending on the power company involved, a request for correction may be not acted upon immediately.

After selecting an antenna site, it is a good idea to check back in a few days to confirm that the channels first observed are reliable. Sometimes a fluke signal will arrive because of unusual atmospheric or weather conditions and will be strictly temporary. In fringe areas, all signals will vary in strength significantly with time of day and season; allowing for this, the best possible site should be chosen to provide reserve signal strength.

In the case of a commercial TV cable company, a proposed antenna site should actually be monitored for several months, in fair and foul weather, before a permanent site commitment is made; unfortunately, this is not always done.

Where the required correction can be provided otherwise, adjacent channels should not be placed on the corrective installation because of the added complications and expense.

The problem with placing channel 3 on the same cable as channel 4, as an example, is that the sound carrier of channel 3 will beat against the video carrier of channel 4 and cause a herringbone interference pattern on the TV screen. Commercial cable companies prevent this interference by reducing the sound carrier about 15 dB below the video carrier level on each channel and/or using expensive filters with special channel processors.

If it becomes necessary to place an adjacent channel on a small system, this can be done by using an aural carrier reducer such as the Jerrold ACR-25 on the lower channel and equalizing the level of the video carriers on these adjacent channels. This will provide satisfactory operation most of the time, but some herringbone may be noticed if the upper channel signal level drops in level as compared to its lower adjacent channel. Automatic gain control (AGC) would prevent this, but it would require individual channel processors or other rather expensive equipment.

It will be noted that in a given metropolitan area the TV channels assigned by the FCC are non-adjacent channels. Very typical assignments are channels 4, 5, 7, 9, 11, and 13 for one area and channels 2, 4, 6, 8, 10, and 12 for another area. This allocation of non-adjacent channels is done to prevent interference between channel signals when receiving TV "off-the-air." Channel locations in the VHF-TV frequency band are shown in Chart 11-1on next page.

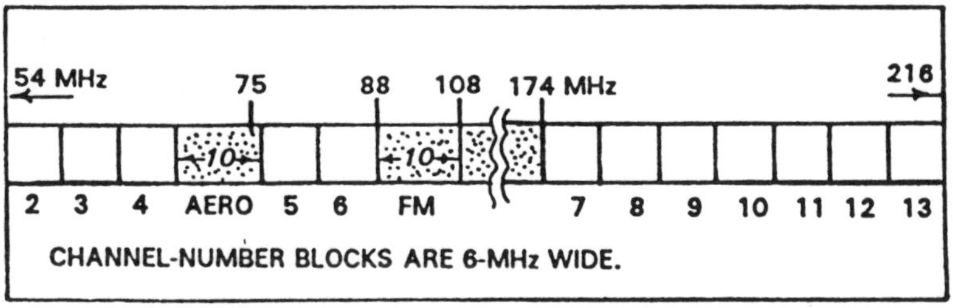

Chart 11-1. VHF TV Channel Distribution.

Antenna installations. A home-owner may buy a competitively-priced antenna, put it up on his roof-top, get fair-to-good reception, and then assume he has the best possible reception for the area. He probably does not know how good television can be with a well-designed installation.

In practically all fringe and suburban areas, multipath (another name for ghosting) is present on some channels. Where the desired channel signals arrive from the same direction, *a high gain, highly directional all-band antenna will minimize this ghosting.* This may mean using an antenna rated "fringe" or "far fringe" in a suburban area. Where the antenna picks up too much signal, an attenuator may be installed to reduce the level. Sometimes single-channel antennas are required in addition to an all-band, where one or more channels arrive from different directions.

When using more than one antenna, it should be remembered that a Yagi antenna designed for a particular channel will also pick up some signals on other channels. This pickup may be from the sides of the antenna or from the back. Channel-blocking filters must be used to keep out these unwanted channels.

Yagi-type antennas that are broad-banded to cover either the channels 2-6 band, or the channels 7-13 band are available and can sometimes be used to advantage.

The Yagi antenna, mentioned frequently in this handbook, and in radio amateur literature is a special antenna configuration designed to receive (or transmit) especially well at a particular frequency. The four antennas seen below in Fig. 11-7 are of the Yagi type.

The bottom antenna shown is a *5-element Yagi* with the elements labeled **R**, *reflector*, **D**, *driven element* and **1**, **2**,and **3** *directors*.

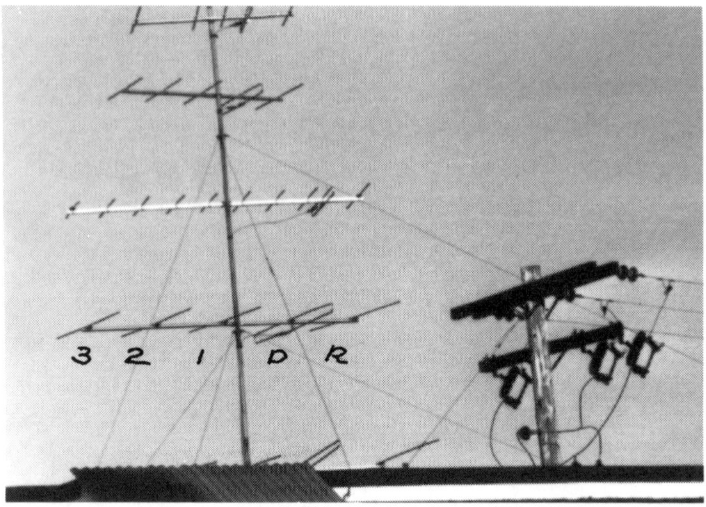

Fig. 11-7. Yagi Antennas Receiving Four Different Frequencies.

In the figure, the driven element **D** is tuned to a particular frequency or channel, and is approximately one-half wavelength of that frequency. For example, if that particular antenna were intended for 2-meters (150 MHz), the length of **D** would be one-half of that wavelength or one meter, about 39 inches.

Increasing the number of directors increases the gain, as well as the directivity of the Yagi.

The antenna is named after Hudetsugu Yagi, of Tokyo University. The name Yagi is often capitalized, but has become so generic that the capitalization is often dropped.

The Yagis pictured are on TV-mast tubing on the roof of a motel. Apparently no thought was give to probable noise occurring from the power-line structure about fifteen feet away.

Fig. 11-8 shows a TVI corrective installation's antenna structure. This uses two far-fringe antennas in a horizontal array, and each pointing in the direction of the TV-stations.

Note that the far-fringe antennas use a configuration of elements evolved from the basic Yagi antenna–however it has many reflectors, driven elements and directors in an arrangement permitting excellent operation at all channels. Here the paired antenna array brings in two low-band channels, three high-band channels and the FM-radio band.

Fig.11-8. Corrective Installation using Horizontally-Stacked All-Channel Antennas.

The horizontal stacking can be done with either all-channel or single-channel type antennas. With such stacking, two identical antennas are operated side-by-side with their outputs fed in phase to a common lead-in cable.

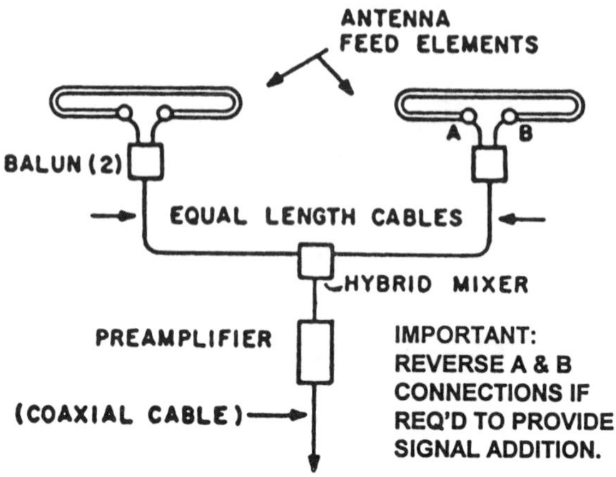

Fig. 11-9. Horizontal Antenna Stacking.

The antennas may be spaced horizontally with the rear elements only one or two inches apart; this is not critical. The antenna booms should be parallel. The spaced antennas are connected together using two identical lengths of coaxial cable feeding a weatherproof hybrid-mixer. (See Fig. 11-9.) The actual length of cable (A) is not too critical but it should be equal in length to cable (B) within 1/4" or so. It is essential that the matching-transformers (baluns) shown in the figure be properly phased. If connecting antennas together results in a signal loss over that of one antenna by itself, the leads to one of the baluns should be reversed. A signal gain will then be noted.. This phasing can be done before the antennas are mounted up on the mast.

When installing an antenna, be aware that foul weather TVI does not radiate from a point source, but from corona plumes all along the conductor. Also, if you are at some distance from the line, the foul-weather noise level usually will peak with your antenna pointing towards a transmission-line tower, because of its elevation, and because of radiation and reradiation effects at the tower.

The head-end 60-Hz power voltage-booster, and the TV mixers, filters and amplifying equipment, required for the corrective installation are shown in Fig. 11-10–enclosed in the metal box mounted on the antenna pole. [30-V AC arrives via the underground TV cable and is supplied by the closest of 3 residences connected].

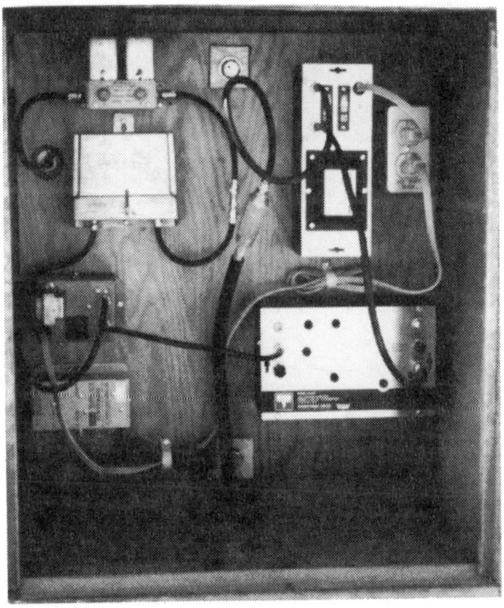

Fig.11-10. The Head-end Equipment Box.

Chapter Twelve

CABLE TV LEAKAGE INTERFERENCE

Introduction

CATV (Community Antenna TV) signal leakage *does* result in interference. Strictly speaking, it is *not* power-line interference, but *it is* RFI and it is important to power-line interference investigators. Such leakage does radiate and reradiate interference from cables and equipment mounted on power company lines, and in so doing is often suspected of being power-line interference. Leakage is a breakout or escape of the television channels and other signals from a CATV system. This occurs at the site of a cable *opening* or *discontinuity*, such as a break in the cable sheath or a poor connection where the signals radiate into the airspace; the opening is usually not visible.

If you are a cable TV manager, engineer or technician, you know why you are concerned about cable TV leakage.

If you are a power company engineer or technician, you are perhaps not so sure why you should be concerned, but you *should* be aware of *any* interference radiating from your power system. This chapter is meant to show you how to identify TV-cable signal leakage on your company's power lines, or on your complainant's TV set. It will help you to distinguish leakage from power-line interference. The chapter will also make you more knowledgeable about cable company procedures in their monitoring and correcting leakage in their cable system.

If you are a TV watcher, you are definitely concerned about TV interference, regardless of who is responsible–your neighbor, the power company or the cable TV company. Most people do not know where the problem is and very often call the power company with their interference problems.

Radiation From Cable Leaks

Many examples of interference radiating and/or reradiating from power lines have been shown in this book as causing interference to radio and/or TV, and this chapter illustrates other examples. Fig.12-1 shows on a television set two examples of interference from *cable leakage,* (A) and (B), one example of *off-the-line* VCR interference (C), and as a review, one example of *power-line* interference (D).

A

B

C

D

Fig.12-1.

A. Adjacent Channels Beating One Against the Other.
B. Two Channel 3's on Same Picture, One with Slightly Offset Video Carrier Frequency.
C. VCR Interference on TV.
D. Power-Line or Other Electrical Noise.

Why Is The Cable TV Company Concerned?

The CATV company is concerned about cable signal-leakage control (1). to comply with FCC regulations, (2). to prevent over-the-air interference to aeronautical navigation and communications,(3). to maintain good public relations and (4). to reduce the number of cable system trouble calls.

Since 1985 cable TV companies have been required by the FCC to meet certain leakage limit requirements. Since July 1, 1990, they have also been required to have in place prescribed programs for the detection, correction, and reporting TV leaks.

Some cable companies, usually small companies, are exempt from these rules, *provided* they do not utilize the frequencies in the aeronautical bands for any of their system TV channels. These companies, nevertheless, must report to the FCC that they are not using aeronautical frequencies.

Why Is The FCC Concerned?

The FCC is concerned about TV cable signal leakage because it is charged with the control of all radiations in the RF spectrum in the U.S. The FCC has issued thousands of licenses to radio stations of many different types to operate in most all of the RF spectrum, and no one has the right to transmit any signals in this spectrum without a license or without permission from the FCC. The "official" RF spectrum in the U.S. is 3 KHz to 300 GHz (gigahertz).

The CATV company often places many TV channels (distant and local) on their cable, but these channels are supposedly confined to traveling within shielded cables for distribution to their customers. A cable system may use any of the channels shown in Table 12-1, page12-16.

They have the permission from she FCC to do this as long as these channel transmissions do not go out into airspace via cable leakage. They have no license to *broadcast* over the air.

The FCC is reasonable about this, however, and does permit

a minimum of leakage as long as the leakage "radiation" does not cause harmful interference and the leaks found are repaired.

Harmful interference is usually interpreted like this: if a party is bothered with interference to the extent that he complains to the power company, to the cable company, or to the FCC, it is harmful interference. There are some qualifications concerning people living in far-fringe areas receiving off-the-air signals considerably outside of radio and TV station service areas. This is where they cannot tolerate even a small amount of interference. These people are *not protected* by the FCC rules–but this does not mean that the FCC or the power company will not cooperate. In the author's experience, they usually will consider special circumstances. An example of a special circumstance might be a party in a far-fringe area who has been receiving good TV (because of an excellent antenna system) for years, and suddenly cannot receive a good signal because of a new, noisy power line, or a leaky TV cable.

2. The FCC is in charge of enforcing *leakage limit* rules which were established because of a number of incidents of interference to aircraft communication from TV cable signal leakage occurring in the years 1976 to 1980.

Limits for TV Cables; Report of Leaks

The FCC considers a level of 50 microvolts-per-meter (50 μV/m) at any one leak location excessive; however, all leaks found to be above 20 μVm must be repaired. If the leaks found are 50 μVm or over, these must be reported and repaired with the date found and the date corrected also reported.

All leaks above 50 μVm are entered into a *Cumulative Leakage Index* (CLI) calculated for the system. The CLI must result in a number less than 64 in order to meet the FCC requirement. The formulas for this calculation, which consider the number of leaks found on the system and their levels, and the percentage of the system monitored (which must be 75% or more and include the oldest portion of the system) are found in Ref. 12. The calculation, however, is usually made by computer.

The portion of the cable-TV spectrum of most concern to the Federal Aviation Administration (FAA) and to the FCC is shown in Fig. 12-2.

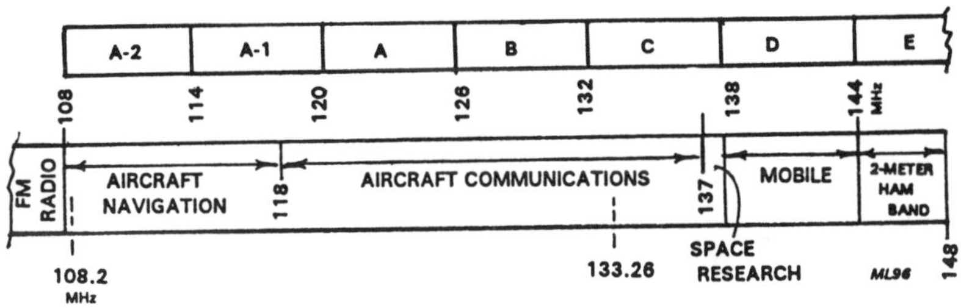

Fig. 12-2. Relation of VHF-Aeronautical Frequencies to Cable TV Frequencies.

Note that the aeronautical "over-the-air" 108 MHz to 137 MHz band is completely used by "on-the-cable" TV channels, which are *meant to stay on the cable.* The 108.2 MHz and 133.26 MHz locations shown in the bottom of the figure are examples of where a signal is often placed by a CATV company on its system for *leak detection monitoring.*

Cable TV Leakage Measurements

The leakage measurements are made at a cable TV frequency selected by the TV cable company (example, 133.26 MHz) using a dipole antenna with its element lengths calibrated for the frequency, and using a calibrated field strength meter indicating, or correctable to, microvolts-per-meter (μVm). For accuracy, the dipole should be held at least 3 meters above ground. The leak measurement should be taken at a nominal 3 meters (10 ft is OK) from the leak, with the dipole antenna mounted on a fiberglass insulated pole Be careful: Do not get the antenna near high voltage. You are connected to this antenna by field strength meter's cable.

Three meters equals 10.17 ft and 10 ft is the nominal distance used by most technicians. Actually, the distance of 10 ft is also the minimum recommended for holding the dipole elements from ground and nearby metal objects, when taking your final reading for your reporting.

It would not do to just go out and measure the level of leakage on a random channel of a TV-system cable to arrive at a leakage number. The level of leakage from the channel you selected might not be as high as from other channels.

It is advisable to have a special standard signal of precisely controlled frequency and level placed on the TV cable for your leakage tests.

Airspace Measurements

Flyovers for cable-TV signal leakage are now the preferred method, and being used by many CATV operators to meet their annual CLI reporting to the FCC. The specialized airborne system used by Flight Trac, Inc. provides an efficient approach to system compliance certification. It also geographically identifies and maps "hot spots" of out-of-compliance leakage, permitting the cable company to quickly pinpoint and correct this leakage. For a block diagram of the system they use, see Fig.12-9, page 12-15. Below is shown a typical aircraft used.

Fig. 12-3. Typical Aircraft Used by Flight Trac [20] for Airborne Measurements.

The CLI measurements per the FCC are preferably taken in the aeronautical VHF band of Fig. 12-2. They may also be taken at other TV frequencies where the level readings are correlated to frequencies in this band.

When a signal is placed on the system for leakage monitoring in an aeronautical band, this signal must be equal in level to the highest adjacent TV channel signal. The monitoring signal is crystal-controlled, and the leakage-detector used is also crystal-controlled to match the transmitter frequency.

It is important to realize that leakage where it escapes from the cable *consists of many frequencies, that is, the TV and FM channels and other signals on the cable.* All of these various leaking frequencies will not be observed at the same level at a leakage site. The TV signal levels placed on the cable are intentionally not equal to start with, to allow for the higher attenuation with distance for the higher frequencies. Metal present at the leak site radiates as an antenna; this radiating metal may be a loose connector, a broken sheath or associated hardware. A particular configuration of metal parts does not radiate equally at all frequencies.

Where Are The Leaks?

We may think of leaks as being on the larger TV feeder cables of ½" diameter or more strung on the power poles, but cable companies report that most leaks occur on the smaller diameter pole-to-house TV service drop cables. On the service drop we may have a break in the sheath, a poor connection at the splitter or directional tap on the pole, a poor connection at the house entry (Fig. 12-4), a broken or punctured cable at a staple location or elsewhere, or poor connections on equipment within a home.

Fig. 12-4 Likely Location for a TV Cable Signal Leak. Poor connection at house-entry splitter.

Leakage Complaint No. 1.

Signal interference from the TV cable to a home not "on the cable" and possibly to aircraft overhead is shown in the photo of Fig. 12-5. (The vertical lines in the center of the picture represent the interference. The diagonal lines are not interference but are a part of the picture; these lines are actually freeway lanes photographed by the Skycam 4, a traffic-watch helicopter, as indicated by the cars on the freeway.)

The power-line RFI investigator, let us call him Don, observed this leak at a house near the power line and reported it to the CATV company after making some preliminary observations. He noted that the TV did not have the symptoms of power-line interference but did have the symptoms of multipath (ghosting).

He observed the following, which led him to conclude that cable leakage was the cause of the problem.

Fig. 12-5. Channel 4 *Leakage* Signal and Channel 4 *Off-the-Air* Signal Received on the Same Antenna.

1. The complainant had an antenna that was pointing towards the TV station and into the power line as well as into the TV cable. He had severe ghosting on channel 4, Seattle.

2. A neighbor a block away, also not on the cable, did *not* have ghosting on channel 4, Seattle. His house was on the other side of the cable, with his antenna looking away from it.

This was sufficient evidence that the problem was TV-cable leakage, but Don also observed, confirming his conclusion:

3. The vertical bars on the TV picture (representing the horizontal blanking and sync pulses required in TV sets for picture scanning) were in the middle of the screen. These vertical bars are normally off of the screen to the left, and appearance in the picture means severe displacement.

It seemed to Don that the severe displacement of the ghost picture to the right could be caused by the extra delay for the TV signal to travel the several miles via the coaxial cable from the CATV head end to the complainant's home.

Don called the CATV company to report the problem, and they sent out a technician to check for TV-cable leakage.

When he arrived on the scene, the cable technician got out his Vitek Tracer I and proceeded to take leakage measurements.

He started in the complainant's driveway at pole 1, as shown below; here the panel meter of the Tracer I showed a low reading of 2 μVm. Moving on to pole 3, it read 30 μVm, and then at pole 4, he read over 50 μVm.

This leak was obviously out of FCC compliance. The cable technician then climbed pole 4 to further explore the problem. Here he found a leaky connection from the splitter to the service-drop cable. He replaced the connector, and this immediately dropped the leakage to zero at that pole. Checking back with the complainant, he found that he had restored ghost-free TV to the complainant's house, and Don's suspicions about cable signal delay, along with a cable leak, causing the ghosting was correct.

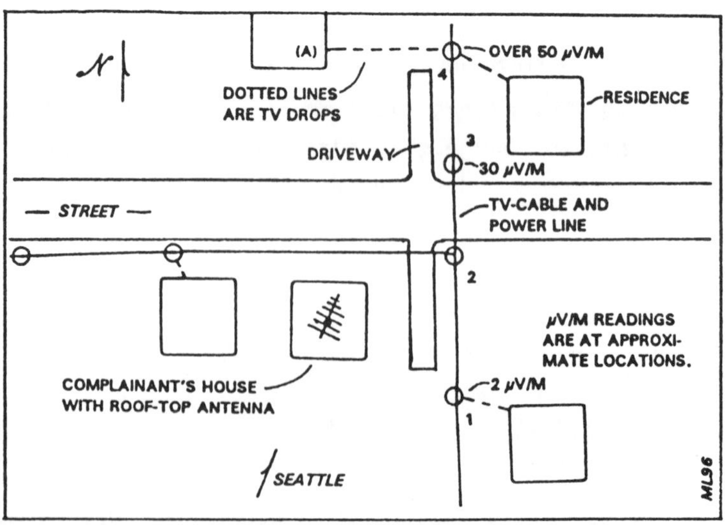

Fig. 12-6. A Cable-TV Leakage Search.

Leakage Complaint No. 2

In this complaint, Don found that at the complainant's house there was objectionable "co-channel" interference on channel 3 and ghosting (multipath) on channels 4, 5, and 7. Further, as a clue to the interference source, two channel 3s were being received; one of these was a weak channel 3, received of-the-air, and another was a weak channel 3 received as a result of TV-cable leakage.

Although the horizontal bars on channel 3 were drifting up across the screen, their spacing was constant. This indicated two steady frequencies quite close together, beating against each other. This would occur with two TV-video carriers trying to occupy the same channel if one of these had a slight frequency-offset relative to the other.

A cable TV technician came out to check the problem. He very shortly found a cable-TV leak near the residence, and after repair, the horizontal bars disappeared on channel 3, and the ghosting cleared up on 4, 5, and 7.

The Leakage Detector

There are numerous makes of leakage detectors on the market, and you will see advertisements and descriptions of most of these by scanning a few recent issues of *Communications Technology* magazine [13].

The leakage detector used by Don in the above two complaints is shown in Fig. 12-7.

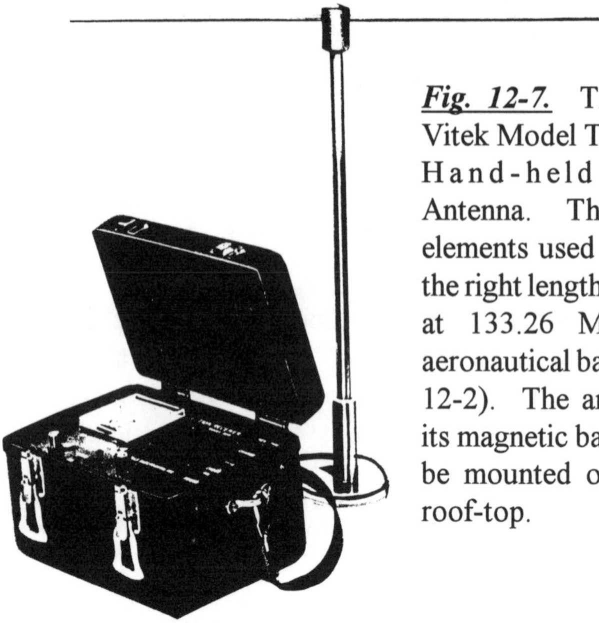

Fig. 12-7. The Portable Vitek Model Tracer I with Hand-held Dipole Antenna. The antenna elements used here are of the right length to resonate at 133.26 MH, in the aeronautical band (see Fig. 12-2). The antenna with its magnetic base may also be mounted on a vehicle roof-top.

The monopole antenna used on the leakage flyover is shown in Fig. 12-8. The whip portion (A) is pulled out and locked at the proper length for the measurement frequency.

Fig. 12-8.

Leakage Measurements with a Field Strength Meter

As a power-line interference investigator you may wish on occasion to measure TV cable leakage, and you can get a good approximation if you construct a simple extendable-element dipole and connect it to your field strength meter (FSM). See Fig. 7-10, Chapter 7. Then, with the elements on your half-wave dipole extended to the correct length for the frequency observed, note the μV reading. Then convert the microvolts to microvolts per meter (μVm), as follows:

$$\mu\text{Vm} = \mu\text{V} \times .021 \times \text{MHz}$$

To start your search you may need a preamplifier between the dipole and the FSM, and the Radio Shack No. 15-1108 may be used with a 9V transistor battery for power. Then as you close in on the leak, you would remove the preamp to make accurate measurement.

At the leak site, make your measurements holding the dipole 10 ft from the leak, orienting the dipole for maximum signal leakage pickup, and keeping your dipole 10 ft away from other conducting objects.

If this is a leak that has been causing a problem to one of your power-company customers, chances are it will be over 50 μVm, and out of compliance with FCC rules. Reporting this to the CATV company gets prompt action.

A Few Facts About Cable Ingress

In addition to signal leakage or egress from a TV cable, we also have noise ingress. CATV companies with two-way data and communications services on their system are very much disturbed by the cable ingress, that is, noise getting *into* the cable system. Ingress may occur at the sites of cable leakage.

This noise can interfere with cable's high-speed on-line access, digital communications, and telephony; it can seriously slow down the operation of data modems.

The two-way channels used are generally between 5-40 MHz in the "sub-VHF" band, with the range of 5-15 MHz being most susceptible to noise resulting from ingress. Critical life-line telephone circuits are being placed in the 20-40 range.

Most noise caused by ingress comes from noisy devices and poor connections within connected homes, and from defects in the TV cable drops to these homes.

The power companies at present are most concerned about *egress*, that is, TV interference to its customers. However, these power companies could also be potential customers of the cable company for data transmission.

Also, a reliable two-way system requires a low noise floor on the cable, which means ingress must be held to a low level.

For an excellent description, a good understanding of RF noise ingress to a cable system, and its control, see the Hewlett-Packard document *Sweep and Ingress* [14].

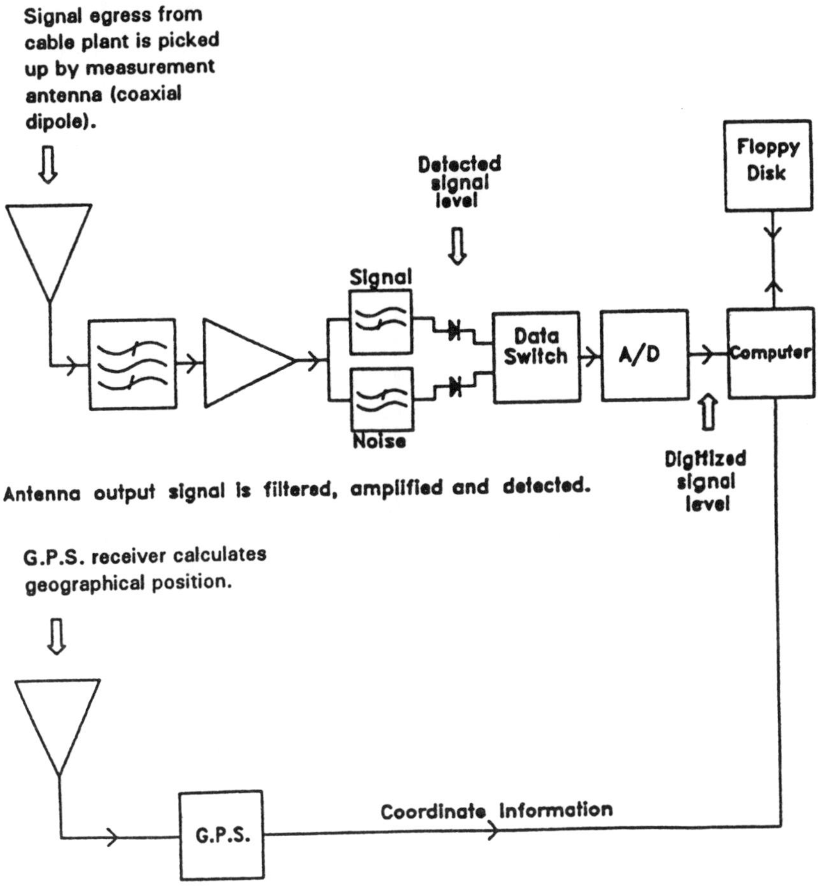

Fig. 12-9. Flight Trac Airborne CLI Measurement System [20].

Cable TV Channels

Table 12-1 shows the various channel frquencies you may find on, and sometimes leaking from, a CATV cable.

Table 12-1. **On-the-Cable TV Channels.**

CHANNEL NO.	VIDEO CARRIER	CHANNEL NO.	VIDEO CARRIER
SUB-VHF		**SUPER-BAND**	
T- 7	7.00	J	217.25
T- 8	13.00	K	223.25
T- 9	19.00	L	229.25
T-10	25.00	M	235.25
T-11	31.00	N	241.25
T-12	37.00	O	247.25
T-13	43.00	P	253.25
		Q	259.25
LO-BAND		R	265.25
2	55.25	S	271.25
3	61.25	T	277.25
4	67.25	U	283.25
5	77.25	V	289.25
6	83.25	W	295.25
MID-BAND		**HYPER-BAND**	
A-2	109.25	AA	301.25
A-1	115.25	BB	307.25
A	121.25	CC	313.25
B	127.25	DD	319.25
C	133.25	EE	325.25
D	139.25	FF	331.25
E	145.25	GG	337.25
F	151.25	HH	343.25
G	157.25	II	349.25
H	163.25	JJ	355.25
I	169.25	KK	361.25
		LL	367.25
HI-BAND		MM	373.25
7	175.25	NN	379.25
8	181.25	OO	385.25
9	187.25	PP	391.25
10	193.25	QQ	397.25
11	199.25		
12	205.25		
13	211.25		

SUMMARY

- Cable TV leakage <u>is</u> RFI.
- Leakage is the breakout or escape of TV signals from TV cables.
- CATV companies are required by the FCC to limit leakage of TV signals from their cables.
- The maximum leak (egress) of TV signal permitted from a particular location is 50 μVm.
- The main concern of the FCC is that CATV leakage does not cause interference in the aeronautical frequency band.
- Excessive leaks must be repaired and reported by the cable company to the FCC.
- Reports to the FCC of leakage on the system are required annually.
- Leakage measurements for the annual report may be made with a properly equipped aircraft.
- A Cumulative Leakage Index (CLI) is calculated by computer where leakage measurements and certain system parameters are entered. The CLI number obtained must be less than 64 for system compliance.
- Leaks occur on the TV cable at a discontinuity in the cable sheath or at a loose or poor connection.
- Cable leaks sometimes result in complaints that may be investigated by either CATV people or power company people.
- Two complaints are described and the clues which led to their resolution by the CATV company are described.
- The concern for cable ingress is described.
- Cable ingress may occur at the sites of cable leaks.

Chapter Thirteen

AC POWER AND TELEPHONE NOISE

This Chapter is based mainly on information found in the excellent telephone seminar handbook, "*Noise Transmission, Inductive Interference*" by Roger Tokarz [1]. It is based mainly on Chapter 4 of that book, and is excerpted and edited here by the "*AC Power Interference*" author for use by power company people. [It contains some basic information that power company engineers and technicians should know about *why and how power lines cause telephone noise.*]

Sentences or phrases in brackets, as well as some of the underlining and italicizing, are added by the "*AC*" author, particularly for power company and other "non-telephone" company people.

Introduction

[In general, to power company people, power-line noise means radio and TV interference originating in arcing or corona sources on a power line, with these sources causing RF-noise radiation from the power-line conductors and/or hardware.]

[On the other hand, to telephone company people, power line noise means noise in telephone lines caused by magnetic induction into such lines. This induction is from the AC power-line magnetic field which contains harmonics of 60 Hz that are heard and identified as *noise* in telephone lines.] The fundamental power-line frequency of 60 Hz itself is not a problem because it is filtered out by telephone equipment.

In this chapter we assume the telephone lines, or wires, consist of multiple pairs in sheathed cables.

The Source of Telephone Noise

Nearly all of the noise heard on a telephone cable pair has its source in the **POWER SYSTEM**, as will be explained. A cable pair refers to the pair of telephone wires bringing telephone messages to your home or business via a cable from the telephone company central office. It is one of many twisted pairs of insulated copper wires bundled in an aluminum-sheathed cable. This cable may be *buried*, strung on *power company distribution-line* poles, or on nearby *telephone-company* poles.

A common misconception, particularly among telephone technicians, is the idea that the earth is a giant receptacle for noise. [And that this “earth” noise is being induced into the overhead telephone cable pairs.] This simply isn’t true. Rather, the *noise* is present in the *huge magnetic field* around the AC power-line conductors, enveloping the telephone cable into which the noise is magnetically induced.

[For radio and TV noise we are concerned with the electric field, not the magnetic.]

The Multi-Grounded Neutral Power System

The most common configuration of AC power distribution systems in the United States and Canada is called the Multi-Grounded Neutral System (MGN). The MGN system is a three-phase distribution system. It uses three phase-conductors and a neutral conductor (usually a size or two smaller than the phase-conductors). Since the pole transformers feeding power to customers from an MGN system are connected to one of the phase wires and to the neutral, it is common practice to carry only one phase wire and the neutral on the poles feeding a group of customers.

[Note that there are no neutral conductors on transmission lines.]

In general, the more innocent a power line looks, the more likely it is to cause problems. We are seldom bothered by the so-called "high lines" for a couple of reasons. The first is, telephone lines seldom parallel *transmission lines* and the second is that transmission lines are not multi-grounded neutral systems. It is the power *distribution lines* that cause us the most trouble.To see how cable pairs are affected by the power system, it is necessary to understand the nature of the magnetic fields surrounding the power system.

To see how telephone cable pairs are affected by the AC power system, it is necessary to understand the nature of the magnetic fields surrounding the power distribution system.

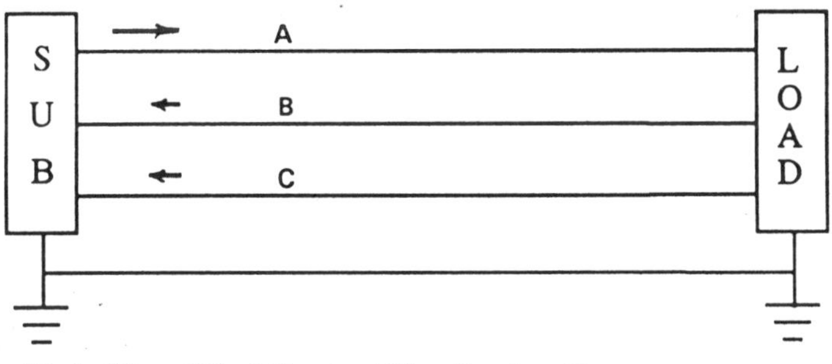

Fig. 13-1. Simplified Power-Distribution System.

As seen in Fig. 13-1, phase A current travels from the substation to the load where it accomplishes its work and then must return to the source. Because of the way the three- phase power-line transformers are wound, the returning current divides in half, with half returning on the phase B wire and half returning on the phase C wire (as indicated by the lengths of the arrows).

Whenever a current flows in a conductor, a magnetic field is formed around the conductor. The *magnitude* of the surrounding magnetic field depends on the magnitude of the current, and the *polarity* of the magnetic field depends on the direction of the current flow.

If we consider all of the conductors which make up the "power line" as a "bundle" of conductors rather than individual conductors, the magnetic field caused by the phase A current traveling toward the load should be exactly canceled by the magnetic fields of the two returning currents. It should be, but it usually is not.

If we now add the B and C power-line phase currents and their return paths, we have the distribution of currents shown in Fig. 13-2.

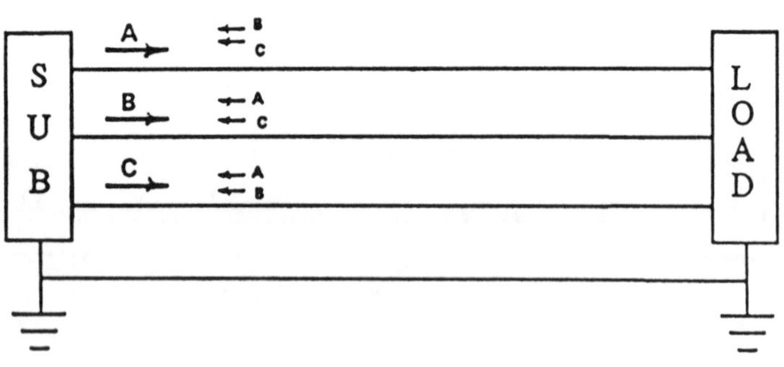

Fig. 13-2. Phase-Current Distribution in Three-Phase Power Line

Of course, all three of the phase AC currents do not travel down the power line simultaneously as shown. Each is separated from the others in time; that is, they are out-of-phase with one another. They are traveling the paths shown, but not simultaneously. In any case, so long as the currents are as shown, [that is, B + C equals A] the net magnetic field around the power line is zero and no current flows in the neutral.

However, power-line phase currents are almost never perfectly balanced in this way. The load consists not of a single unit as shown in Fig. 13-2 but many small loads distributed along the power line. Add to this the fact that it is impossible to predict which loads will be "on" and which will be "off" and we have a problem. [Remember that we are talking about 3-phase *distribution* systems–the systems that cause us the telephone noise].

Simply stated, the problem is this: whenever the phase currents are not equal (almost always), the *algebraic difference* between these currents must return to the substation in some path other than the phase wires. At first thought this doesn't seem to be a big problem. After all, isn't there a neutral conductor in the same bundle, and couldn't the return current take this path? And, if the current returned in the neutral, wouldn't the magnetic field around the neutral tend to cancel some of the fields in the phase wires? Yes, yes, and yes. All of these things are true, but at the same time there are factors which cause part of this current to take an unexpected path.

Ground Return Currents

The unbalanced current attempting to return to the substation has two possible paths...the neutral conductor and...the earth under the power line. It might seem that an electrical current would always take a metallic conductor path rather than a path through the earth, but the rules of current flow prevail. Current will take the path of least resistance. In the case of a power line, the *resistance of the neutral conductor wire* increases with distance *much* faster than the resistance of the earth so, in general, the longer the power line, the more current will return via the earth.

The current that returns in the earth (called ground return

current) is not a single current but many tiny currents distributed over a large cross-section of earth. Because of this *wide disbursement*, this ground return current is unable to create any effective canceling [of the magnetic field] from currents in the power line. The result is *a net magnetic field around the power line.*

The *net* magnetic field around the power line is the source of nearly all of the interfering noise found on cable pairs, so it is important to have a clear understanding of how this field behaves. If we had the ability to "see" magnetic fields, we might see a power line as shown in Fig. 13-3.

Fig. 13-3. The Power-Line Magnetic Field.

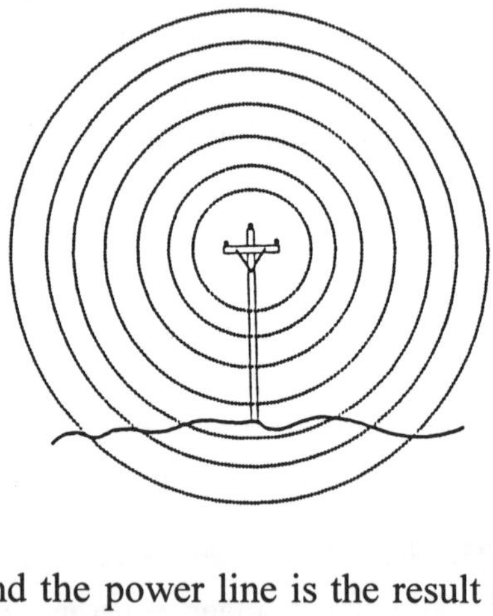

The magnetic field around the power line is the result of *alternating* current, so it is a *moving* field, expanding and contracting in a pattern several hundred feet in diameter. Notice too, *the field does not stop at the surface of the soil but actually penetrates the earth and affects buried telephone cables as well as aerial cables.*

One of the rules of electric induction says, "Whenever there is relative motion between a conductor and a magnetic field, a voltage will be induced in the conductor." Because the magnetic field around the power line is a *moving field* (caused by its 60-Hz alternations) every conducting object in the field will have an AC voltage induced into it.

You can test this concept quite easily in the field, near a power distribution line. Connect one lead of a telephone handset to ground (a screwdriver driven into the earth will do) and connect the other lead to a strand of barbed wire on a fence. Be sure it is on wood posts. The noise you hear will sound the same, except for its level, as the sound you will hear when connected to one wire of a cable pair under the same power line. [The 60-Hz fundamental frequency is filtered out in the handset and this you do not hear].

> Two of the factors that determine the magnitude of the noise voltage induced into a particular telephone cable are the strength of the power-line magnetic field and the distance the telephone cable parallels the power line. *A few spans of paralleling usually presents no problem, while several miles often means trouble.*

The Theory of Magnetic Shielding

To understand how these power-line magnetic fields affect telephone cables, it is necessary to dissect a cable and perform a bit of "stop action" magic. Fig. 13-4 is a *representation* of the basics of a cable with only a single wire inside the outer metallic sheath. The sheath is shown as a series of aluminum tubes. The sheath would be continuous aluminum tubing except that a section is removed at each location where we want to make connections or otherwise get at the pairs. If the cable is properly installed, each opening in the aluminum sheath is bridged across using a substantial conductor (normally a #6 stranded copper wire) and two compression clamps. We assume, in Fig. 13-4, that the sheath forms a continuous *electrical* path from one end of the cable routing to the other end.

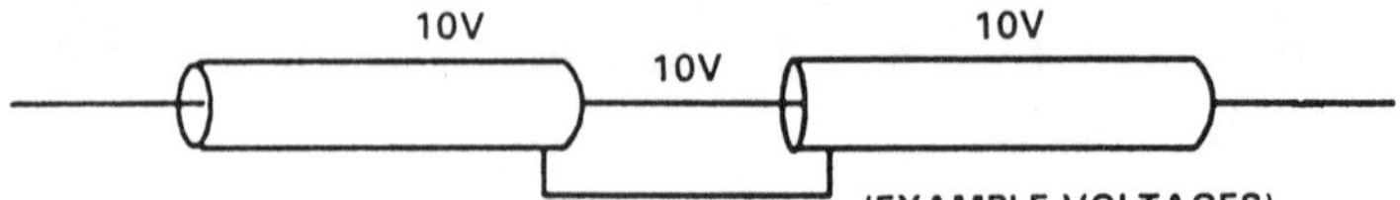

Fig. 13-4. Induced Voltages from the Power-Line Magnetic Field.

For this discussion we will assume this cable is within the magnetic field of an AC power line and the magnetic field is [expanding and contracting, as it moves through the telephone cables.]

Remember, the rules of induction state that whenever there is relative motion between a magnetic field and a conductor there will be a voltage induced into the conductor. Here we have *two conductors*: the aluminum sheath and the copper wire inside, both of which are exposed to a moving magnetic field.

It might seem that the sheath would somehow protect the wire inside from this moving field, but it does not.

> To the magnetic lines of force, the sheath is merely a *transparent piece of aluminum*. The magnetic lines *penetrate* the sheath and induce a voltage into the sheath *and into the wire inside* just as though the sheath were not there.

Because the sheath-conductor and the inner-conductor are substantially the same distance from the center of the power-line field, and both conductors are the same length, it is logical to assume they will have the same induced voltage. This is true and is so indicated by the lengths of the arrows in [Fig. 13-1].

Since both of the conductors are open (they have no connection on either end) there can be no current flow. Remember, the magnetic field around the power line is a result of alternating current, so the voltages represented in Fig. 13-4 are AC voltages. These are represented in a "stop action" fashion to make the explanation easier.

Fig. 13-5 reflects the cable sheath as it should be, with grounds on each end. Now there is a path for the induced current to flow along the sheath, to ground and back to the other end of the sheath. This current return path is shown by the dotted line "S."

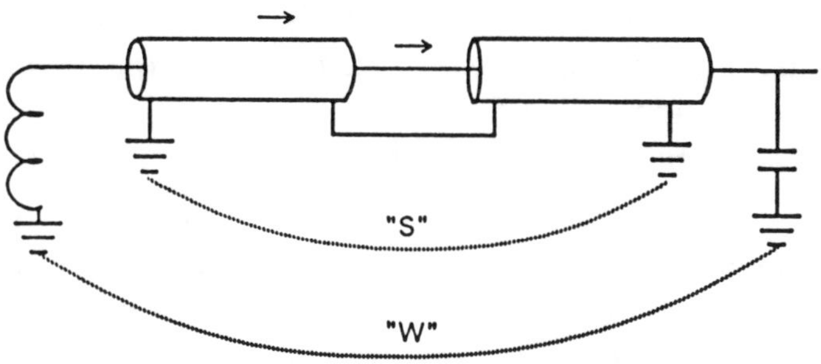

Fig. 13-5. Induced Current Flow, Telephone Cable,.

Also shown are the paths to ground on the single wire. The coil on the left-hand end is typical of the path to ground found in the telephone company central office on *each wire of a telephone pair,* and the ground return path is indicated by path "W".

On the right-hand end of the single wire a capacitor is shown to ground. This is not a single capacitor as is shown. What really exists along this and every other wire in the cable is a capacitive relationship to the grounded sheath and all other conductors in the cable. Think of this capacitance to ground as thousands of tiny capacitors distributed all along the wire. Each capacitor allows some of the AC to leave the wire and return to the central office end.

How much current flows in either of these two paths depends on the impedance (AC resistance) of the paths. Of the two, the sheath is much the better path. A typical value for sheath resistance is 1/2 ohm per 1000 ft. The connections across sheath openings and to ground should be #6 copper wire. Because of these low resistances, sheath currents can be quite large. Values between **1** and **10** Amps. are considered normal.

The wire inside the sheath is another matter. [This wire] itself is of much smaller gauge than the sheath which means its resistance is much greater. Added to this is the resistance of the central office ground path (typically 200 ohms) and the "resistance of the capacitance-to-ground path" (perhaps 2000 Ohms) depending on a number of factors. In any case, the induced current in the single wire inside will be very small.

Now for the magic. Another rule of induction states that **"whenever a current is caused to flow as a result of a magnetic field, that current will create its own magnetic field which is 180 degrees out of phase with the field which caused it."** Quite a mouthful and probably not worth memorizing, but it does explain how the sheath works. The large current flowing in the cable sheath creates a magnetic field which is exactly 180 degrees out of phase with the power line magnetic field.

The wire inside the sheath now has two magnetic fields to contend with ... one from the power line and one from the sheath. If, from the point of view of the wire inside, these two fields were equal and opposite, no noise current would flow in the wire. That is, the sheath would be able to exactly cancel the effect of the power line.

It is important to recognize that this shielding action is a direct result of sheath current. This means that the bonding wires must be in place across every sheath opening, and there must be connections to ground on each end of the sheath.

Pollution on the Telephone Pair

It would be wonderful if we could count on the sheath to exactly cancel everything the power line induced into our cable pair, but it just doesn't happen that way. Theoretically, it is possible to achieve perfect shielding at some frequencies. This is the exception and not the rule. The rule is, no matter how good a job we do of bonding and grounding there are always *more induced noise currents* there [from the power line] than the sheath can cancel. It is these leftover currents we are most interested in because they are the currents which add up and cause noise. Fig. 13-6 shows the distribution of these currents on a telephone pair.

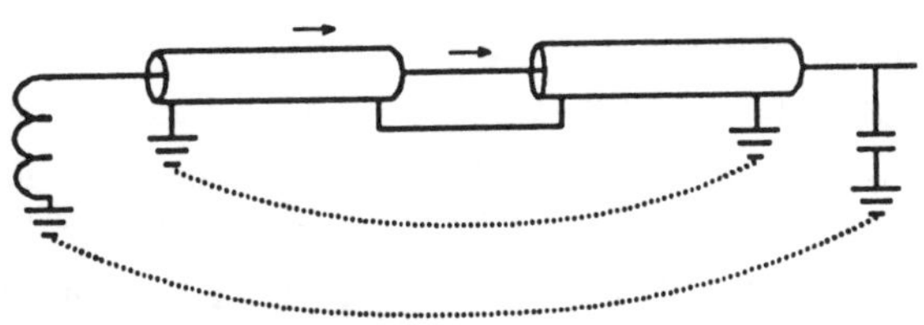

Fig. 13-6. Paths to Ground, Telephone Pair.

Fig. 13-6 is an approximation of the AC paths to ground that exist on a typical telephone pair. As the arrows in the drawing indicate, the leftover currents circulate along each of the wires of the pair seeking paths to ground. The capacitive relationship between each wire and ground provide these paths. Because these currents travel in the same direction on both conductors of the pair at the same time they are called "longitudinal currents."

These *unwanted* longitudinal currents may be thought of as the environment in which the telephone circuit must operate. They are a form of *electrical pollution.*

These currents are not a new discovery by any means. For many years they were considered the province of telephone engineers and noise experts. This was partly because test equipment suitable for measuring these currents was too expensive to be supplied to all technicians and partly because not everyone thought the technicians needed to know. The test equipment limitation has been solved long since. As for technicians needing to know, opinions vary among companies and managers. It is clear, however, that a knowledge of the behavior of these currents is essential for trouble-shooting noise problems.

The present interest in these currents brought about a number of changes, one of which was the name. These currents are now known as Power Influence (PI) although many meters with Ng scales are still in use. While "power influence" is a more descriptive name of these currents, "noise to ground" provides a better mental picture of how they are measured. Since the currents are seeking paths to ground, we need only give them such a path for their measurement through the meter. Fig. 13-7 illustrates how this is done. Note that the ring and tip sides are connected together for this measurement.

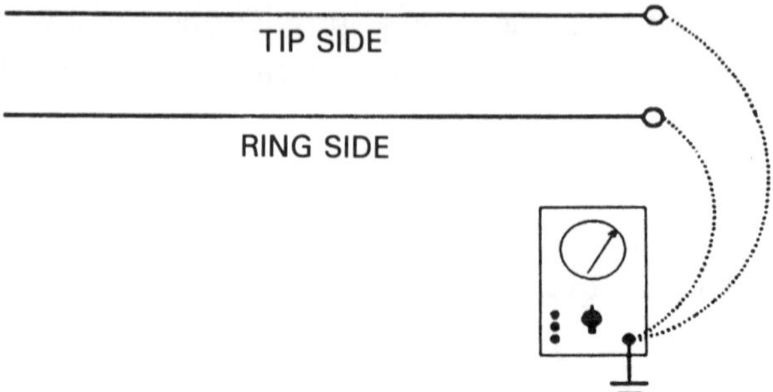

Fig. 13-7. Measuring Power Influence (Noise-to-Ground).

[***Regarding the tip and ring.*** In earlier days when switching was done by operators in the central office, telephone cords were used to plug into the various jacks on a switchboard in order to complete connections. The 3-wire telephone cord used a special plug on its end with a telephone-circuit pair connected to the tip and ring and a ground wire connected to the sleeve, as shown in Fig. 13-8.

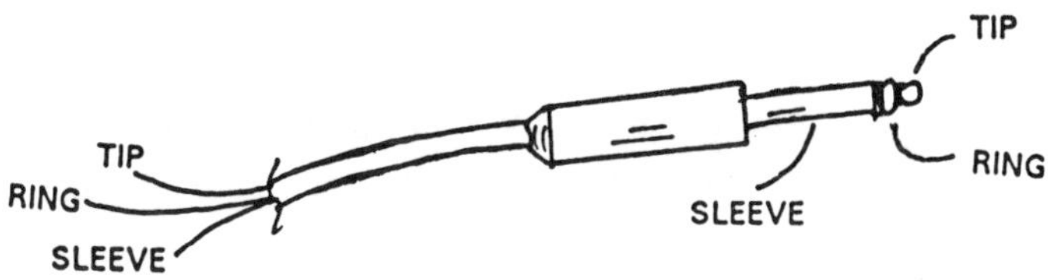

Fig. 13-8. Telephone Plug with Tip, Ring and Sleeve.

The wire of the pair connected to the tip of the plug became known as the tip-conductor, or just the "tip," and the wire connected to the ring became known as the "ring." The wires of the pair are also sometimes called the tip side and the ring side of the circuit.]

Quality of the Telephone Circuit

Assuming the power influence currents were induced equally on the tip and the ring (the two wires of a telephone pair) and the sheath treated each of the wires equally in its canceling action, it would seem that the two currents would be exactly equal.

The capacitance paths (which the power influence currents take to ground) must be as close to equal as possible. This is one of the important reasons the pairs are twisted within the cable. The twist allows the relationship between the tip conductor and all other conductors in the cable to be the same as the relationship between the ring conductor and all other conductors.

But it isn't only the wires of the pair themselves we have to worry about. Every device in the circuit must look electrically identical to tip and ring wires including the central office and the station equipment. Given all of these requirements, it's not surprising we have been unable to achieve perfection.

If the PI paths to ground within the cable were all equal, and if the devices in the circuit all looked electrically equal to the tip and ring wires, we would have perfect balance and a very high quality circuit.

Perfect or not, there is a term which describes how well the telephone circuit is designed and maintained. **THE TERM IS BALANCE.** The definition of **BALANCE** is: a number which represents the **QUALITY** of the telephone circuit.

To meet the minimum quality standard, the "balance" must be at least 60 dBrnc* of power influence. The formula is *Balance = Power Influence minus Circuit Noise,* and if the balance is below 60, a physical problem is indicated such as a split pair, a grounded carbon, a wet cable, a bad connection or even a fault in the central office or in station equipment. The EMC [EMI] engineer might properly say Balance = Common Mode Rejection [see next chapter, 14].

*dBrnc refers to dB above reference noise level with a C-Message filter in the circuit. The reference noise level is −90 dbm; that is, 90 dB below one milliwatt. This reference was determined in the telephone industry as the level where no noise could be heard. This is the zero noise level, and anything measured as being below this is simply called 0 level. This reference avoids the need to use negative numbers when measuring power influence and circuit noise. The C-Message filter effectively blocks the 60-Hz fundamental frequency, but passes voice frequencies readily.

Whenever a frequency is generated (even the low power-line frequency of 60 Hz), other frequencies are also generated unless great care is taken. These extra frequencies are exact multiples, or harmonics, of the generated frequency. There are times when these extra frequencies are desirable. In music, for example, if an unseen musician plays a tune, first on a banjo and then on a guitar, we can easily recognize which instrument is which, not because of the notes but because of the extra frequencies generated with the notes. The same is true of the human voice. Two singers singing the same notes sound different to us because of the unique patterns of harmonics produced by the vocal cords.

With three-phase power systems the harmonics present in the 60-Hz waveforms are mostly the odd multiples, since the even multiples tend to cancel themselves out in the *three-phase* transformer winding, and thus no even harmonics are present. The signal induced into the sheath and pairs of the telephone cable consists of 60 Hz plus 180, 300, 420, 540, 660, 780, 900, 1020 and 1140 Hz, and sometimes even higher order odd harmonics. *Unfortunately these are all* voice frequencies [which we can readily hear].

Most of the harmonics contained within the 60-Hz waveform travel in the power system, as does the 60 Hz. That is, the harmonics travel along the three-phase conductors 120 degrees apart and so have the natural canceling between the phase wires as was shown in Fig. 13-2. The *odd-triples,* such as 180, 540, 900 and 1500 Hz, however, are another matter. These harmonics come up in-phase on all three phase-wires and so are forced to return to the source either via the neutral or through the ground-return path. These frequencies then are more likely to create power influence than the other harmonics, and the ninth harmonic of 60 Hz (540 Hz) is especially troublesome.

Harmonic Sources

Why are these harmonics present in what's supposed to be a 60-Hz system? Well, some are present at the output of every generator. In a well-designed generator the harmonic output is minimal; however, *over-excitation of a transformer generates harmonics.* This is a common occurrence where a power customer has a heavy inductive motor load and capacitors are used to improve the power factor. When the customer shuts off the motors the voltage goes high and causes the transformer to produce harmonics. Also, a capacitor bank with one fuse blown can create a mass of harmonic energy. Unfortunately, capacitors may also provide a low impedance path for ground-return of harmonics. Also, rectifiers on the line such as at aluminum plants can cause serious harmonics. Of course, if the 60 Hz present was a pure sine wave we would have no harmonics.

Even though the 60-Hz fundamental frequency is present in the magnetic field surrounding the telephone cable, it is filtered out in the telephone handset and is not heard.

Non-power-company harmonics can arise from sources such as from SCRs in customer equipment, and produce even and odd harmonics of 60 Hz. these may be conducted into the power line. Here the "evens" as well as the "odds" may have a noise level of well over 60 dBrnc and can cause a real problem

SOME COMPONENTS OF POWER INFLUENCE

[This section includes excerpts from Chapter 9 of Roger Tokarz's *Noise, Transmission, Inductive Interference*, third edition, and personal communications with Roger].

1. The total noise of concern from a power line includes the fundamental 60-Hz frequency, and its many harmonics up to and through 3000 Hz.

2. The "C Message" filter in the telephone handsets subtracts the 60 Hz noise; thus the total telephone noise includes only the harmonics of 60 Hz.
3. If we had a pure 60 Hz to deal with, that is, a pure sine wave (remaining pure), we would have no 60-Hz harmonics.
4. At the generator site the output is close to being a pure sine wave, but it is always distorted, often severely, along the way by power utility usage and other problems (such as over-excited transformers).
5. Many, many harmonics result. One thing in our favor is that the even-order harmonics, such as 120, 240, 360 etc., are self-canceling by transformer action in the a three-phase power system. But the odd harmonics, such as 180, 300 420, etc. give us much trouble (these harmonics are odd because they have multiplication factors of 60 and an odd number).
6. There is not a single magnetic field around the power-line conductor, but many such fields–one for 60 Hz and one for each of its harmonics.
7. The sheath really does operate as described earlier, but the sheath is trying to cancel the effects of 60 Hz and the effects of each individual harmonic. So, there is not just one induced voltage on sheath and pairs. There are individual induced voltages for 60 Hz and for each of the harmonics present. In a similar manner, there are several individual sheath currents, each trying to cancel either 60 Hz or one of the harmonics.
8. Sheath current does a much better job of canceling some frequencies than it does others. The rule is: the higher the frequency, the better job the sheath does. At 540 Hz the sheath is expected to cancel about 50% and at 1020 Hz the sheath cancels about 80%.

9. Because of the C-Message filter we can say: power influence and circuit noise consist of harmonics, those harmonics that can get through the C Message filter. The frequencies between about 300 and 3000 Hz do get through and these are the culprits.
10. It is the ground return current, which represents the unbalance in the power lead, and thus the magnitude of the magnetic field around that lead. If this current were a perfect sine-wave it would contain only pure 60 Hz and would produce no telephone noise (because of the filter in the telephone handset).
11. Five factors which contribute to power influence are: (1) The strength of the magnetic field around the power line.

    (2) The distortion of the 60-Hz sine wave.

    (3) The position of the telephone cable with relation to the center-line of the power-line and its magnetic field.

    (4) The length of the parallel between the cable and the power line, and

    (5)The efficiency with which the sheath cancels the current induced by the power line.
12. One or more of the above factors will have to be changed if we are to reduce the power influence.

Only factor No. (5), if deficient, can be improved by the telephone company.

The other factors should best be referred to a noise mitigation specialist trained to work with the power company.

Conclusion

The **ODD HARMONICS** of 60 Hz occurring at voice frequencies, and originating on the **POWER SYSTEM** represent **POWER INFLUENCE** and are the dominant cause of **NOISE IN THE TELEPHONE SYSTEM..... POWER**

INFLUENCE *reduction* may require the assistance of the power company in order to mitigate a noise problem.

Reference 1. *Noise, transmission, Inductive Interference.* Roger Tokarz, 1991, Third Edition.* This book of fourteen chapters is recommended for further study. ***Much interesting and informative material is presented in this book:***

Bonding and grounding practices....Grounding near power substations (take care!)....Capacitor banks....Buried power cables.

Measuring sheath resistance....The isolation transformer... The induction neutralizing transformer....Customer equipment as harmonic sources....The troublesome 9th harmonic....Power-line resonance at a harmonic frequency.... Trouble shooting with the Power Influence meter.

More on balance....Effects of intermediate grounds on telephone cables....Noise from the customers point of view....

The need for DC on the cable pairs. The telephone receiver circuit....The telephone circuit protector....The cable pair as a capacitor, and adjusting this capacitance.

The meaning of cable-pair loading....Trouble-shooting with the Noise meter and the Termination Test Set....Solving bonding and grounding problems.

Sheath current where a cable is subject to several power-line exposures....use of the clamp-on ammeter in bonding and grounding...Using a single-line drawing for trouble-shooting....Analog and digital transmission on telephone pairs...And more!

****Mr. Tokarz's book*** is available from TAD Consulting, 11611Teepee Road, Spokane, WA 99204-9522, ($18.00 plus $6 S&H). This is a 120-page well-illustrated book in an 8-1/2" x 11" format, and is spiral bound for convenient study. Call or fax 1-509-244-3111 for information.]

Chapter Fourteen

POWER DISTURBANCES TO COMPUTERIZED EQUIPMENT

This chapter by Daryl Gerke, "Power Disturbances to Computerized Equipment," *is used with permission of the copyright owners, R & B Enterprises of 20 Clipper Road, West Conshohocken, PA 19428-2721, publishers of the annual "Interference Technology Engineers" Master (ITEM).*

Disturbance Modes and Types, Definitions, Mitigation.

Power disturbances are increasingly an EMI problem. Computerized systems are particularly susceptible to the noise and voltage fluctuations on power lines. Definitions, applicable documents and mitigation techniques are discussed.

By Daryl Gerke, P.E., Kimmel Gerke Associates, Ltd., St. Paul, MN.

INTRODUCTION

For commercial and industrial computer and microprocessor based systems, interference problems generally fall into one of several categories: regulatory compliance (FCC/VDE/VCCI); radio frequency interference (RFI) due to high level electromagnetic fields; electrostatic discharge (ESD); or power line disturbances.

In too many cases, power disturbances are overlooked or ignored by the manufacturer in the hope that someone else -- the user or the power company -- will solve the problem. This is unfortunate, since interference problems related to power disturbances are increasing as modern electronic systems incorporate embedded computers, microprocessors, and other complex solid-state devices. By their nature, these devices operate at low energy levels and high speeds, making them particularly susceptible to power line noise.

The primary responsibility to prevent and cure these problems is with the equipment designer and manufacturer, with a secondary responsibility with the installer or user. Blame can easily be placed on the power utilities and demands can be made that they provide "clean power," but the power company is usually not the culprit, nor does it have much control over the situation. In one recent study, over 85 percent of the power disturbances (such as impulses, surges and sags) were caused by customers, not the utilities. Simply stated, designers and EMC engineers need to address the problem first. Hopefully, this article can help.

DISTURBANCE MODES AND TYPES

One of the reasons power disturbances are difficult is the many variables involved. First, noise can be present -- either common mode or differential mode -- and secondly, disturbances can vary from short spikes to complete loss of power. Compounding the problem is a lack of clear, concise definitions with which everyone agrees. Nevertheless, some definitions will be presented for the purpose of this article.

Common Mode vs Differential Mode

The first crucial definitions regard the "direction" of the noise or disturbance on the power lines, as shown in Figure 1. Noise currents that all flow in the same direction (voltages in phase) are referred to as "common mode" or "longitudinal mode." Noise currents that flow in opposite directions (voltages out of phase with each other) are referred to as "differential mode" or "normal mode."

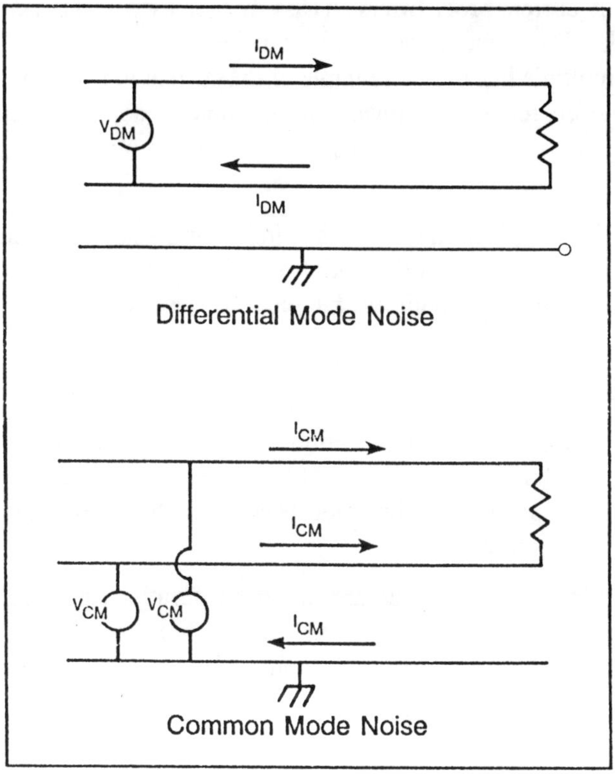

Figure 1. Common Mode vs Differential Mode.

The mode often gives a clue to the source and/or coupling path. For example, differential mode noise suggests that the disturbance is on the same power circuity, while common mode noise suggests that the noise was coupled into the circuit by radiation or crosstalk.

A special case of common mode noise, where the noise is common mode on the phase conductors but is differential with respect to the safety ground, is often seen in switching mode power supplies. It is the result of parasitic capacitive coupling to the cabinet, and is a common cause of FCC and VDE compliance failures. It can also cause problems to electronic equipment sharing the same power circuit.

A current probe offers an easy method by which to determine the mode. The probe is connected around each line individually, then around multiple lines. For example, if the current increases with two lines, the currents are flowing in common mode; if the currents decrease to zero with two lines, the current flow is in differential mode.

This distinction on modes is important, since common mode fixes may not work on differential mode problems, and vice versa. Both modes should be addressed; to do otherwise is to solve only half the problem.

Disturbance Types

The second set of definitions addresses the various types of disturbances on the power lines. At the present time, the terminology for power disturbances has not been formalized, but there are common terms in wide use.

Most classifications are based on duration and levels, and include voltage variations (sags, surges, out-ages), frequency variations, waveform distortion (voltage and current), continuous noise of "hash," and transients.

Table I provides an overview of these disturbances and their most likely sources. Usually the source is not utility related, but due to nearby motors, radio transmitters, or non-linear devices such as switching mode power supplies, or variable speed drives.

Effects on Computerized Equipment

Fortunately, not all of these disturbances cause problems for computerized equipment. In fact, only two are very serious threats -- short-term spikes and long-term sags. Spikes can be coupled into the system by crosstalk or even directly through the supply, causing false triggering of internal logic circuits. Sags can cause loss of memory or other volatile data. Other disturbances, such as minor over-voltages, harmonic distortion, or frequency deviations, are usually ignored by computerized systems.

Types of Disturbance	Typical Origin
Sags 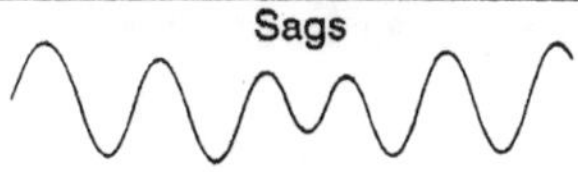	• Lightning • Turn-On of Heavy Loads • Brownouts
Surges 	• Sudden Load Decreases • Incorrect Transformer Taps
Outages 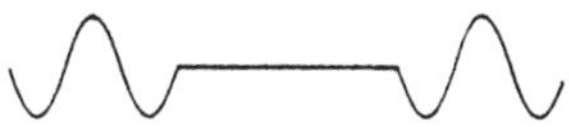	• Severe Weather • Transformer Failures • General Failures
Harmonic Distortion 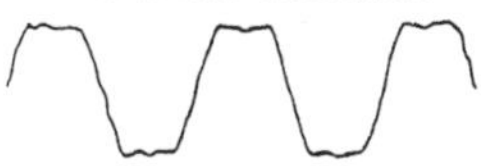	• Rectifier Loads • Switching Power Supplies • Variable Speed Drives
Frequency Deviations 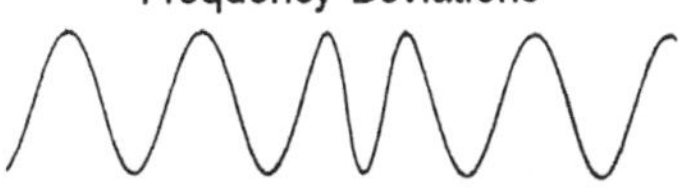	• Generator Instabilities • Regional Network Problems
Transients 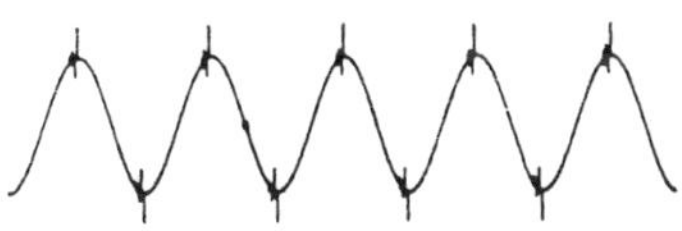	• Lightning • Power Line Feeder Switching • Power Factor Capacitor Switching • Turn-Off of Heavy Motors
Electrical Noise 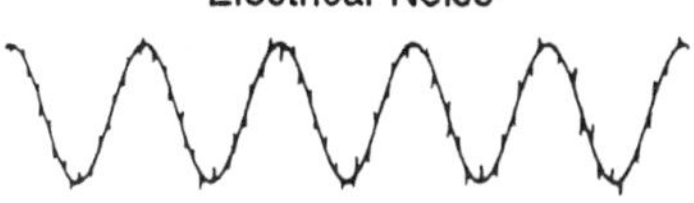	• Radar, Radio Signals • Arcing Utility and Industrial Equipment • Converters and Inverters

Table I. Power Disturbance Types.

Several studies have been made to determine the effects of power disturbances on computers. Three comprehensive and widely quoted studies were by IBM, AT&T and the U.S. Navy. Figure 2 summarizes two of those studies, Allen and Segal of IBM (1974) and Goldstein and Sperenze of AT&T (1982). While the IBM study shows most of the problems due to the transients, and the AT&T study shows most due to voltage sags, these discrepancies can be explained by different test thresholds. The third study, by Thomas

Key of the U.S. Navy, summarized computer failures over a ten year period; undervoltages were the primary power line failure mode.

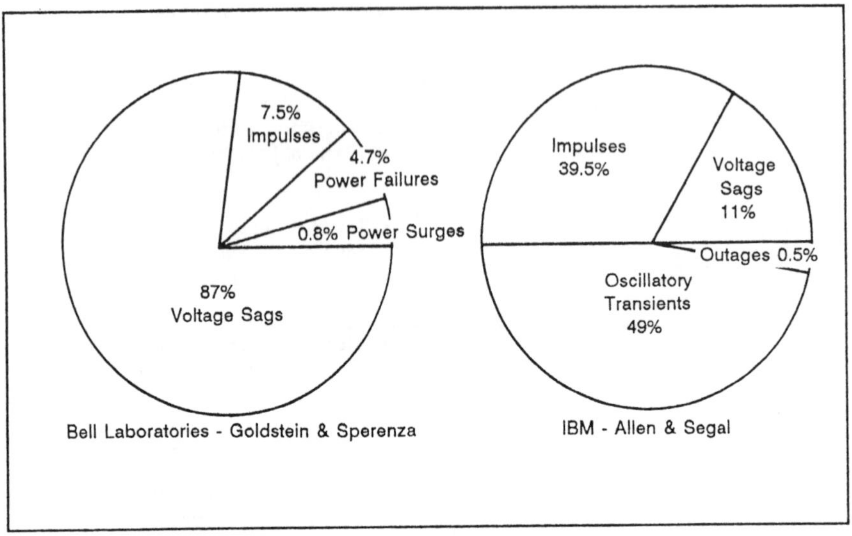

Fig. 2. Results of Studies on the Effect of Power Disturbances on Computers.

The key message is that long undervoltages/sags, and short transients/spikes are the principal power line disturbance failure modes for computerized equipment.

Harmonic Distortion: An Emerging Problem

Although computers are not generally affected by harmonic distortion, they are a major cause of this problem. The problem is not due to the computers or microprocessors themselves, but rather the power supplies. In this case, the victims are often adjacent motors, transformers, and even facility wiring.

Power supplies can represent "non-linear" loads to the power distribution system. Rather than taking current over the entire sine wave, the power supplies take a gulp of current at the peak of the cycle. This non-sinusoidal current (Figure 3) results in current harmonics, as well as voltage harmonics, if the current is high enough to distort the voltage waveform. These current harmonics can cause overheating of transformers, motors, and even neutral lines on three-phase systems. The latter is due to the addition, rather than cancellation, of selected harmonics in the neutral.

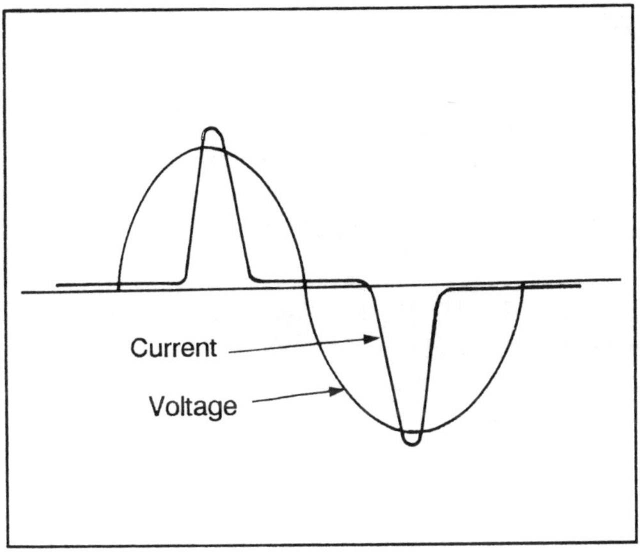

Figure 3. Example of Current Distortion Due to a Non-Linear Load.

If only a few non-linear loads exist in a system, the effects are negligible. In some facilities, however, over half of the loads may be non-linear due to computers and variable speed motor drives. In recognition of this problem, the 1990 National Electric Code now requires larger neutrals in distribution systems supplying non-linear loads. In addition, facilities designers are also using larger transformers and wire sizes for the same reason. In a sense, computers are "getting even" for all the noise they've been subjected to in the past.

GUIDELINES

Although there are no mandated regulations for power disturbances other than the relatively low level FCC/VDE conducted emissions limits, there are industry guidelines and specifications that address this area. In addition, most large computer companies today have internal power specifications to address this issue. Smaller computer companies, and manufacturers of related equipment like industrial controls would be well advised to adopt similar internal standards.

MIL-STD-461

As the EMI standard for equipment procured for military applications, both emissions and susceptibility on the power lines are addressed in MIL-STD-461. Since this is a military standard, however, the applicability to the commercial and industrial environments is somewhat limited.

One MIL-STD-461 test limit that is common in commercial/industrial specifications is method CSO6, the conducted spike test. This pulse, shown in Figure 4, is considered to be a good example of actual power transients, so it has been adopted by many as a test standard. Also, test equipment is readily available to perform this test.

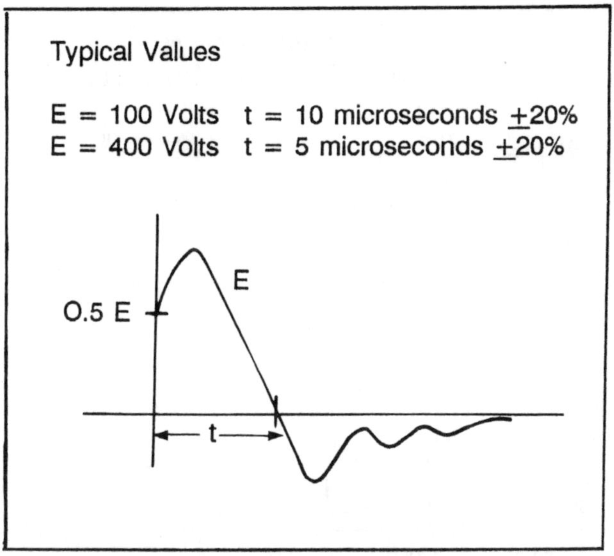

Figure 4. MIL-STD-461 Spike - Method CS06.

"IEEE Recommended Practice for Emergency and Standby Power," (also known as the "Orange Book" gives recommended power limits for computer manufacturers. This curve is shown in Figure 5, and has been adopted by many manufacturers. Computer equipment should operate with no power upsets within these limits. Note should be taken that it assumes a long-term sag of 13 percent, and a total short-term loss of voltage for up to 1/2 cycle. Spikes increase with shorter duration.

This curve also appears in FIPS PUB 94, "Federal Information Processing Standard Guidelines on Electrical Power for ADP Installations," an excellent guide for installing computer systems This curve is often referred to as the CBEMA curve, after the Computer Business Equipment Manufacturers Association.

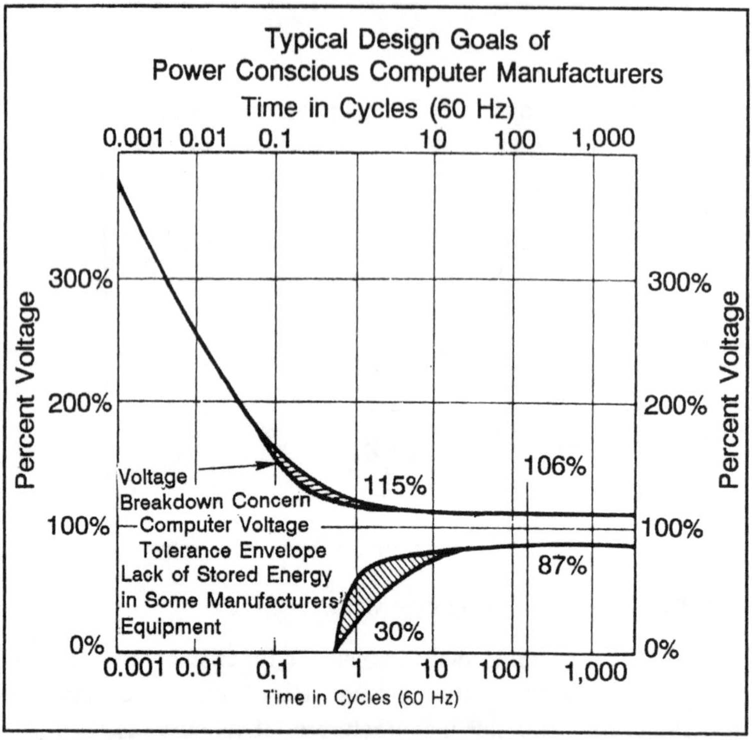

Figure 5. ANSI/IEEE-STD 446-1987 Recommended Power Limits.

ANSI/STD c62.41-1980

This document, "IEEE Guide for Surge Voltages in Low Voltage AC Power Circuits," is essentially a lightning damage specification. Formerly known as IEEE STD 587, it defines surge voltages and currents of up to 6000 volts and/or 3000 amps, and is based on empirical data gathered over many years.

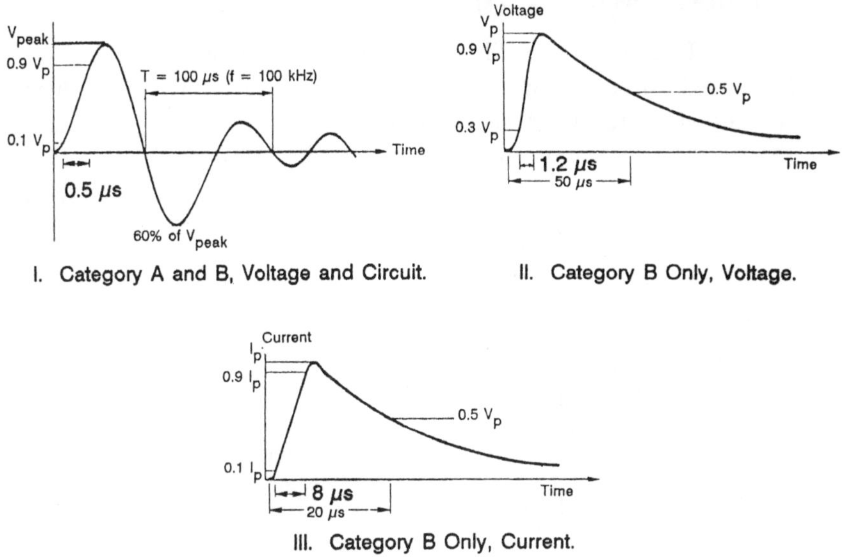

Location Category	Waveform	Amplitude
A - Inside Bldg., Greater Than 10 M from Entrance	0.5 μs - 100 kHz (I)	6 kV 200 A
B - Inside Bldg., Less Than 20 M from Entrance	1.2 x 50 μs (II) 8 x 20 μs (III)	6 kV 3 kA
	0.5 μs - 100 kHz (I)	6 kV 500 A

Figure 6. ANSI/IEEE C62.41-1980 Recommended Transient Waveshapes.

Fig. 6 shows these limits. Category A refers to equipment located more than 10 meters from the service entrance; Category B refers to any equipment within a building. Unless one can guarantee the location of equipment, however, the more stringent Class B limits should be used. These limits are considered damage levels, rather than upset levels. It is probably unrealistic to expect computer equipment to operate through these transients without an upset; the equipment should survive, however, without damage.

ANSI/IEEE STD 519-1981

This document, "IEEE Guide for Harmonic Control and Reactive Compensation of Static Power Converters," gives guidelines for allowable levels of harmonic distortion. It is often used as a reference by power utilities and specifically states that it is not an EMI specification. Nevertheless, it is an important document regarding power line disturbance.

MITIGATION TECHNIQUES

There is a wide variety of devices and products available that can help to mitigate power line disturbances. These range from simple single component spike suppressors to complex automatic on-line uninterruptible power systems (UPS). In fact, a multi-million dollar annual business has grown up in recent years to address these problems.

The proper application of these products is as important as the proper selection. Improper application is a very common cause of power disturbance problems.

In addition, both coupling modes, common and differential, must be addressed. To only protect for one mode is to solve only half the problem, be-

Dedicated Circuits

Transient Suppressors
- Crowbar - Arc Gap Devices
- Clamp - Zener Devices

Filters

Isolation Transformers
- Unshielded
- Faraday Shielded

Voltage Regulators
- Electronic
- Mechanical
- Ferroresonant

Power Conditioners
- Multipurpose

UPS/SPS Systems
- On-line vs Off-line
- Electronic with Batteries
- Motor Generators

Table 2. Power Disturbance Mitigation Techniques.

cause to predict in advance which mode will predominate is difficult, and often both modes are present at the same time.

Ranging from simple to complex (and inexpensive to expensive) the devices and products which can help control power disturbance problems are summarized in Table 2.

Dedicated Circuits

A dedicated circuit is a single circuit with a single load. This is a simple, inexpensive, yet highly effective technique. It works best if the dedicated circuit originates at the main entrance panel, rather than a sub-feeder panel, since this provides the lowest common impedance.

As a variation, a single circuit feeding multiple pieces of computer equipment, with no "noisy" loads like motors or other machinery, can be used. The neutrals and ground wires must be properly sized, since most computers use switching mode power supplies.

Transient Suppressors

These devices divert energy above predetermined voltage levels. Included in this category are gas discharge tubes, metal oxide varistors (MOVs), and zener diodes. Some suppressors are hybrids of the above.

There are two classes of transient suppressors -- crowbars and clamps. The crowbar devices provide a short circuit above the threshold, while the clamp simply limits the voltage excursion at the threshold. The differences are shown in Figure 7.

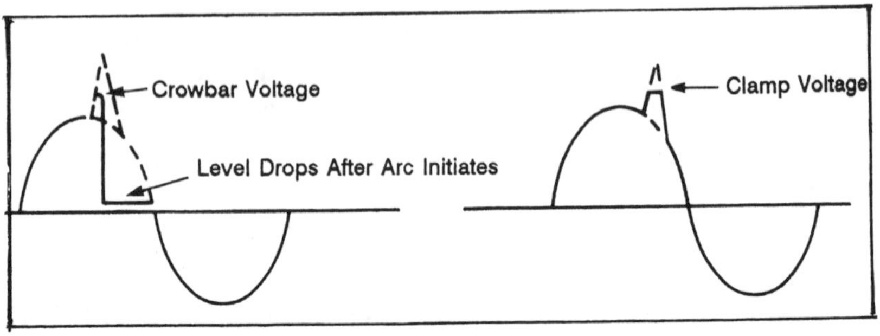

Figure 7. Crowbar vs. Clamp Transient Suppression.

The crowbar suppressors are usually arc gap devices, such as gas discharge tubes. When the gas ionizes, the voltage across the gap drops to a low level, and most of the energy is diverted. The response time of these devices is relatively low, so they do not work on fast transients. They are very well suited, however, for lightning transients.

The clamp suppressors are either zenker diodes or metal oxide varistors (MOVs). These devices act by limiting the voltage at a preset level, and diverting the remaining energy. Since much of the energy is dissipated by the clamp, they are usually rated in joules of energy. A word of caution -- if this rating is exceeded, the device can be destroyed. The response time of these devices is much faster than the crowbar suppressors, with the zenker diodes being faster than the MOVs.

Transient suppressors can be installed for either common mode (line-to-ground) or differential mode (line-to-line) transients. To protect for both modes, protectors should be installed line-to-line, and line-to-ground.

Finally, the effectiveness of any transient suppressor depends on proper installation. Lead length should be kept as short as possible; transients behave like high frequencies, and lead length inductance can severely limit their performance.

Filters

Unlike transient suppressors, which are non-linear devices, power line filters are linear circuit elements combined to pass power frequencies and reject higher frequencies that contain electrical noise. They are most useful for removing lower levels of continuous energy, although they will reduce transients as well. (Since they are linear, however, filters reduce transients proportionally, and do not clamp to a fixed value like a transient suppressor.)

Most commercial power filters today address common mode and differential mode noise. Figure 8 shows a typical filter, having a common mode choke and common mode capacitors (line-to-ground) combined with differential mode chokes and differential mode capacitors (line-to-line).

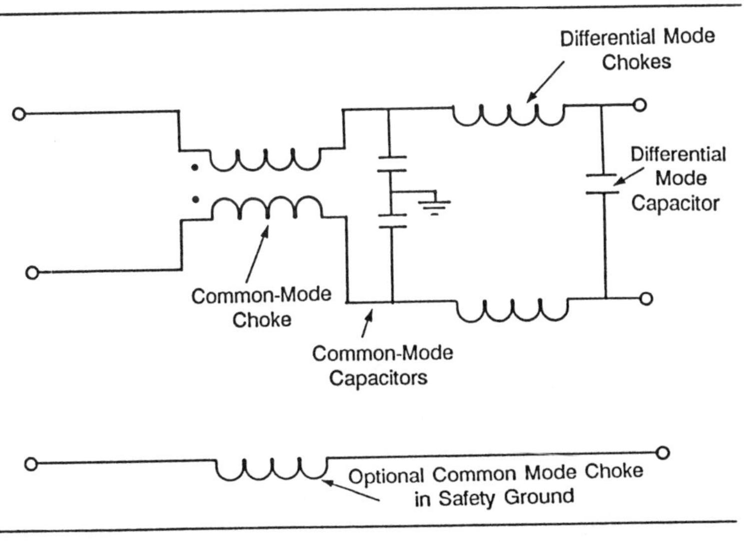

Figure 8. Common Mode and Differential Filter.

A common trend today is to add a small choke in the safety ground lead, which helps with common mode filtering. This can be quite effective against common mode transients, such as lightning. This value must be kept small enough, however, to not interfere with the fault clearing capabilities at 60 Hz.

The effectiveness of filters depends on proper installation. At higher frequencies, filters must be installed in the cabinet bulkhead to prevent coupling around the filter. Also, the input and output wiring must be separated, or crosstalk can couple noise around the filter.

Finally, grounding of the filter is very important. Figure 9 illustrates the effects of poor grounding on a pi-network filter. In this case, the input and output capacitors actually bypass the inductor at high frequencies. Direct contact is best; even an inch of wire at high frequencies results in too much inductance.

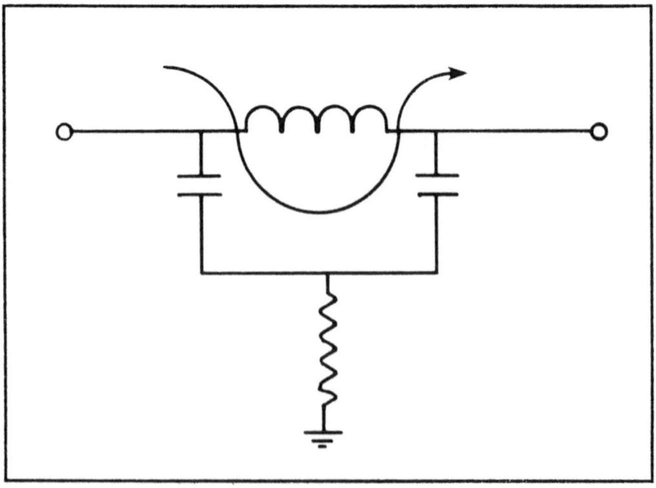

Figure 9. Poor Grounding of Filter.

Isolation Transformers

Isolation transformers are power transformers with no direct current connection between the primary and secondary. As a result, they provide isolation between input and output circuits, as well as a level of common mode shielding.

This isolation and shielding degrades with increasing frequency due to interwinding capacitance. To extend this operation to high frequencies, capacitive or "Faraday" shields are installed between the winding.

These shields can provide common mode or differential mode protection, depending on where the shield is terminated, as shown in Figure 10. For common mode, the shield should be connected to the ground, while for differential mode the shield should be connected to the neutral. The object is to provide a high frequency path for the noise.

If shielding is needed for both modes, two shields can be used -- one for common mode, and one for differential mode. As with filters, installation is important. The shield connections must be as short and direct as possible, or the shield will degrade at higher frequencies.

Isolation transformers work best at lower frequencies (under 1 MHz). Thus for full frequency protection, filters may be needed as well. In fact, the combination of filters and isolation transformers offers a wide range of frequency protection and component sizes with reasonable cost

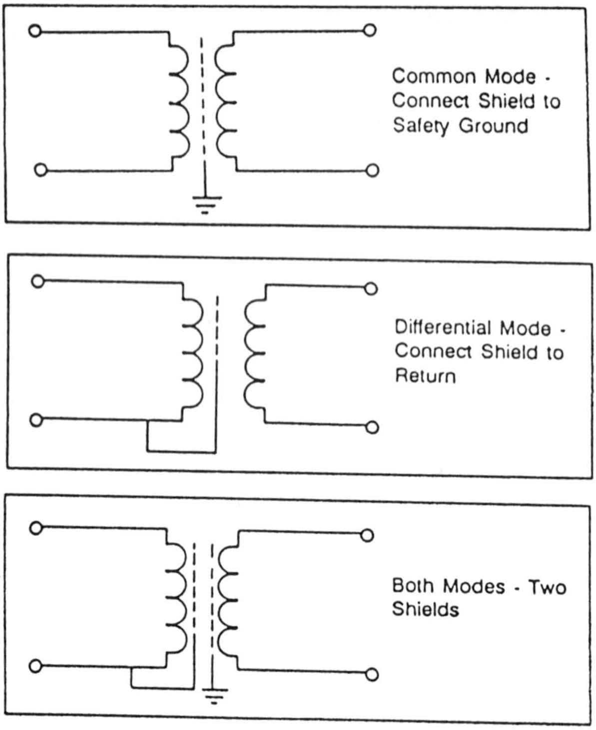

Figure 10. Common Mode vs Differential Mode– Transformers.

Voltage Regulators

Standard isolation transformers do not offer protection against long-term voltage surges or sags. For these types of disturbances, some method of voltage regulation is needed.

A regulator's function is to maintain output voltage within a limited range with varying input conditions. Regulators can be active, using electronic circuits, or passive, such as autotransformers or ferroresonant transformers. Since computers are particularly susceptible to low voltage conditions, regulators are often needed to assure proper operation.

Ferroresonant regulators have enjoyed great popularity due to their simplicity and high reliability. Essentially a transformer on the verge of saturation, they exhibit small output voltage variations with large input voltage variations. They may cause problems, however, when used with switching mode power supplies, since ferroresonant transformers have a relatively large output impedance and switching power supplies need a low impedance. Ferroresonant regulators should be specifically designed for use with switching supplies.

Power Line Conditioners

Power line conditioners are usually used for products that provide several functions in one unit -- isolation, voltage regulation, and noise suppression. They generally do not provide protection against total power losses.

Power line conditioners can vary from hybrid transformer-filter-regulator units to motor-generator sets. They are often marketed in a package for computer operation, and are usually more cost-effective than buying individual protective devices.

UPS/SPS Systems

The ultimate in power protection is the uninterruptible power system (UPS), or its cousin, the standby power system (SPS). The distinction between the two is that the UPS is on-line, and continues to provide power without interruption to the load, while the SPS is off-line, and must be switched into the load circuit.

UPS/SPS systems typically employ batteries for short-term energy storage, and may employ gasoline or diesel generators for longer term energy needs. A UPS/SPS is the only way to protect critical loads for outages in excess of about 0.5 seconds.

Installation of UPS/SPS systems is very important. The system must be isolated from the utility power wiring. Otherwise, utility workers could be harmed, believing they were working on unenergized wiring. Installations should be performed and checked only by competent electricians and engineers who understand the safety issues. IEEE-STD-446, "IEEE Recommended Practice for Emergency and Standby Power" is an excellent guide in this area.

SUMMARY

Power disturbances are increasing as an EMI problem, particularly with modern electronic systems. Computerized systems are very vulnerable to short-term voltage spikes and long-term voltage sags. An emerging problem is harmonic distortion, due to the proliferation of non-linear loads such as switching mode power supplies.

The primary responsibility for preventing and fixing these problems lies with the designer and user. There is little most utilities can do to help, since most problems are generated from within the user's facility. Most utilities, however, are willing to give guidance, and some even have "power quality" engineering groups dedicated to helping with these problems.

While industry guidelines for power disturbances do exist, at the present mandatory requirements do not, other than the relatively low level FCC or VDE limits. Conscientious companies should incorporate internal test standards for their own equipment.

A wide range of mitigation techniques and products are available to prevent and solve power disturbance problems. Proper selection and installation, however, are crucial for success. Both coupling modes (common and differential), must be addressed, and short connections are mandatory for high frequency protection.

BIBLIOGRAPHY

ANSI/IEEE Standard 446-1987, "IEEE Recommended Practice for Emergency and Standby Power," (The Orange Book), New York, NY: IEEE.

ANSI/IEEE Standard C62.41, "IEEE Guide for Surge Voltages on Low Voltage AC Power Circuits," New York, NY.

ANSI/IEEE Standard 519-1081, "IEEE Guide for Harmonic Control and Reactive Compensation for Static Power Converters," New York, NY: IEEE.

"Filters and Power Conditioning," Gainesville, VA: Interference Control Technologies, Inc., 1988.

FIPS PUB 94, "Guide on Electrical Power for ADP Installations," Gaithersburg, MD: U.S. Department of Commerce.

"Power Quality Reference Guide," Toronto: Ontario Hydro, 1989.

DARYL GERKE is an EMI consulting engineer, a practical trouble-shooter and an owner with Bill Kimmel of Kimmel-Gerke Associates, Ltd. of St Paul, MN and Mesa AZ, phone (888) EMI-GURU (364-4878). Bill and Daryl are the authors of an excellent book, *EMI Suppression Handbook.* In this book they describe many practical ways to suppress EMI radiation & conduction in equipment of new design, as well as equipment found in the field. It would be useful to radio amateurs who like to build equipment and are concerned about EMI radiating from such. You may order this handbook from Seven Mountains Scientific at (814)426-6559 ($25 +shipping).

[1] M. O. Loftness. *Power Line Interference - A Practical Handbook.* National Rural Electric Power Association (NRECA). 43301 Wilson Blvd., Arlington, VA 22203-1860. 1992.

[2] W. Janischewskyj, A. M. Hussein, N. H. C. Santiago, *Performance and Analysis of a Microgap Discharge Circuit,* IEEE Trans. on PA&S.

[3] Frank Warburton, formerly New England Power System. Fig. 1-18, photo from *A Practical Handbook for the Location, Prevention and Correction of Television Interference from Overhead Power Lines,* by M. O. Loftness. Bonneville Power Administration (BPA) Division of Laboratories, PO Box 491, Vancouver, WA USA 98666. 1977.

[4] W. E. Pakala, Vern Chartier, *Radio Noise Measurements on Overhead Power Lines from 2.4 to 800 kV.* IEEE Trans. Power Apparatus and Systems, vol. PAS-90, May-June, 1971.

[5] M. O. Loftness, *A Practical Handbook for Correction of Radio Interference from Overhead Power Lines and Facilities.* BPA, Division of Laboratories, PO Box 491, Vancouver, WA, USA 98666. 1980.

[6] Ron Furrer, Umatilla Electric Co-op, Hermiston, OR 97838-3148.

[7] M. O. Loftness, *A Practical Handbook for the Location, Prevention and Correction of Television Interference from Overhead Power Lines,* Bonneville Power Administration (BPA) Division of Laboratories, PO Box 491, Vancouver, WA USA 98666. 1977.

[10] Vern Chartier. IEEE Tutorial course, *The Location, Correction, and Prevention of RI and TVI Sources from Overhead Power Lines.* vlchartier@IEEE.org.

[11] C. F. Clark and M. O. Loftness, *"Some Observations of Foul Weather Television Interference."* IEEE Transactions Paper No. 70TP 104-PWR, Feb. 1970.

[12] Gordon and Beu, *Monitoring and Measuring Signal Leakage, A Service Technician's Guide.* Orion Technical Training Institute, 1003 K St. NW, Suite 825, Wash., D.C. 1990.

[13] Communications Technology Magazine. 1201 Seven Locks Road, Suite 300, Rockville, MD 20854

[14] Hewlett-Packard, Direct Marketing, PO Box 580599-MS 51L-SK, Santa Clara, CA 95052-8059.

[20] Manufacturers, Representatives, Consultants:

Donn Wales & Associates, Inc. Marysville, WA (360)659-5224. For power-line hardware. Hastings & Kearney Products.

Radar Engineers, 9535 NE Colfax St.,
Portland, OR 97220. 1-800-426-3851

Flight-Trac is now Mar-Tech, Jacksonville, FL.
904-720-0082. www.martechengineering.net.

Bearings, Inc., Seattle, WA for Dow Corning #41 Grease
(206) 223-52255

Safety Line, Inc., PO Box 2540
Oakland, CA 94614-0540 (415) 562-4945

[30] Photo Credits:

Fig. 2-4, Kent Mackebon (BPA)

Fig. 5-3, Safety Line, Inc. Oakland, CA

Fig. 6-1, ICOM America, Bellevue, WA

Fig. 6-2, Ron Furrer (Umatilla Electric) Hermiston, OR

Fig. 9-1, Actors are Victor Hernandez, Brian Bozarth, Mark Scribner, and Joe Garazzo, (City of Riverside, CA)

Acknowledgments

I wish to acknowledge the many readers who have encouraged me to prepare a second edition of *AC Power Interference,* and to you I submit this revised version for your study and, hopefully, your enjoyment.

I could not have accomplished this edition, of course, without the careful editing and reworking of the manuscript by my wife, Nancy.

Marv Loftness

Appendix

Supplemental Information to the text and figures shown throughout the book is included here. It opens with a discussion of negative corona. The next few pages will discuss oscilloscopes, opening with a discussion of Hitachi V-509 and the Leader 321 models. Unfortunately these CRT scopes are no longer being manufactured, however the author is aware that these *are* being used by RFI investigators, and also that people have recently obtained them on the used market. It is well worth your search for one of these "beautiful" instruments.

Negative corona. After discussing RFI from positive corona plumes, the author has been asked, "Why don't we have corona noise on both positive and negative half-cycles of the 60-Hz voltage, as we do with sparking noise?" Actually, we do have, but it is very weak.

For an explanation we must turn to the corona physicists, and in doing so I have freely adapted the following from *"Characteristics of RI & TVI Sources"* by W. Janischewskyj[2]:

Weak negative corona does occur during the one-half cycle when the over-stressed power-line conductor is negative, and when, during this time, electrons are repelled from it and move from a region of high electric stress into a more lowly-stressed region. However, slow-moving retained positive ions form a space charge which strengthens the electric field in front of the conductor (the region between the conductor and the positive space charge), but weakens the field in that space away from the conductor (beyond the region where the space charge resides). This effect of the space charge is to confine the negative corona to the vicinity of the conductor (a few millimeters out) and cause little RFI. Then as the positive space charge "slowly" approaches the conductor (on the time scale of tens of nanoseconds), the distance out from the conductor of the electrically over-stressed region reduces below that required to sustain the ionization, and the small corona plumes are extinguished.

After reading the above explanation you may think it is better to just accept (rather than to thoroughly understand) that negative corona, being very weak, is of interest, but of no particular concern, to the RFI investigator.

In the meantime you may want to purchase the excellent performing scope, the very reasonably priced Kenwood CS 4135 (see

Fig. 4-6, Chapter 4). Its main disadvantage is that it requires a 12V DC-to-120V AC converter. This must have at least 11V DC input or it will "drop out;" however, the CS 4135 consumes 40 watts of power.

The Kenwood has an accelerating potential on its cathode-ray tube (CRT) of 12 kV! [Your scope should have at least 4 kV of accelerating potential for adequate brightness in the field].

The Hitachi V-509 Oscilloscope. Before the front panel controls are set, the side-panel controls settings should be made as in Fig. 16-1.

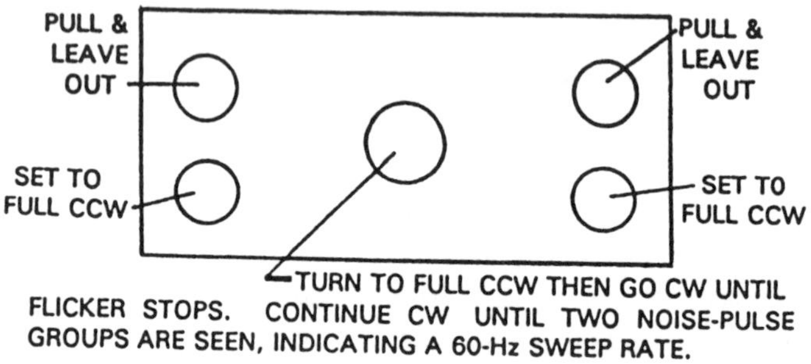

Fig. 16-1. Hitachi V-509 Side-Panel Control Settings.

On the next page, the front-panel control settings are shown in Fig. 16-2. These show the preliminary adjustments to be made when the scope is to be used for power-line noise locating. One exception is that when used with the Scope Controller of Fig. 7-5 (Chapter 7), the variable volts setting should be set at 10 mV rather than 0.1 V.

With the settings shown, you will have a minimum of adjustments to make during your locating procedure.

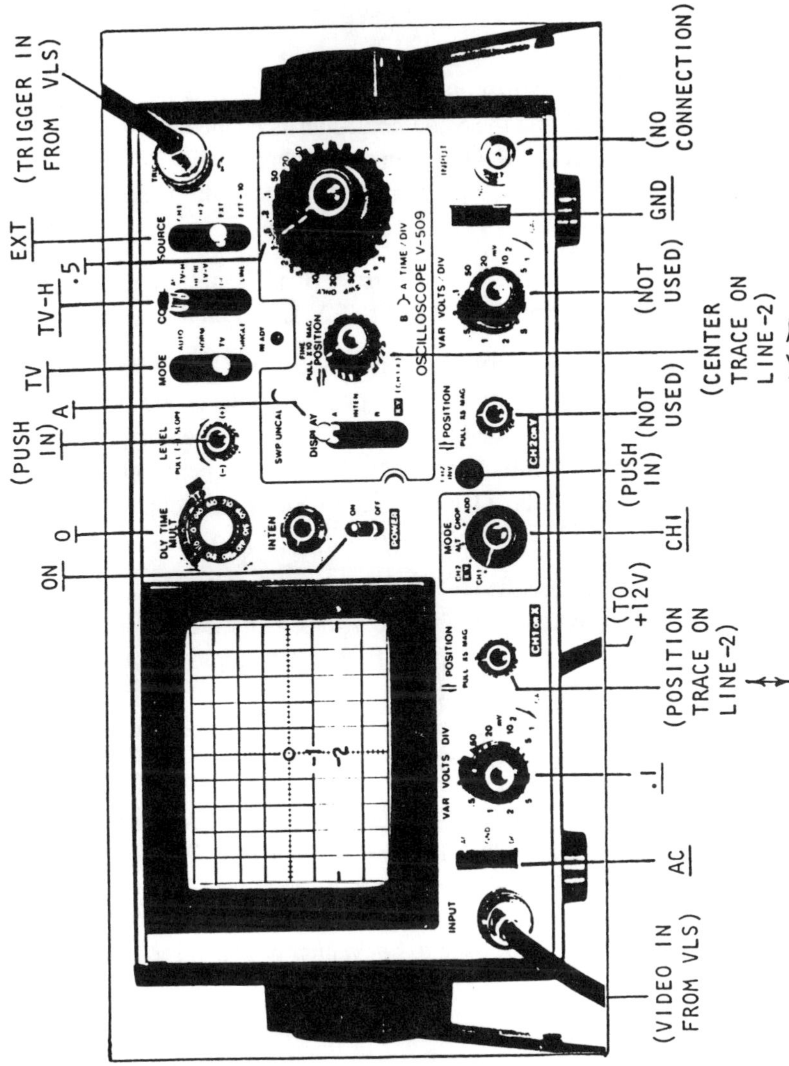

Fig 16-2. The Hitachi V509 Oscilloscope with Control Settings Recommended *at the Start* for the Instrument.

The Leader 321 Oscilloscope. After the oscilloscope settings are made per Fig. 16-3 and the SC-3 controller is connected (with its top BNC cable to EXT TRIG IN on scope back, and its side cable connected to CH 1 on scope front), make sure the scope's TRIG control is *in* and set at its "O" (center position).

On your FSM, tune in the video carrier of a TV station, recognizing it by its "60-Hz" sound, and peak this TV signal on the meter.

With the scope turned on and with the SC-3 VIDEO gain control turned up, you will observe a video pattern on the scope screen (see Fig. 4-9A). Then with the SC-3 BRIGHTNESS switch off, rotate the controller's SYNC control to turn on the 60-Hz sync generator; then rotate it further until the TRIG'D green LED on the scope comes on. You then see the TV video pattern drifting to the right.

On the scope, rotate the small outer knob of the TIME/DIVISION control CCW to where you observe a flicker in the scope pattern on the screen, and then CSW until the flicker just stops. Now when you tune the FSM away from the TV signal, and into electrical noise reception, you will see two noise-pulse groups per one sweep of the scope screen, and you are synchronized at 60 Hz.

A vinyl grommet should be installed as a "thumb-grip"over the horizontal centering control, as was indicated for the vertical centering control on the scope in Fig. 16-3. A larger 5/16" ID grommet is installed on the outer knob of the TIME/DIVISION control. You will then be able to conveniently adjust these knobs.

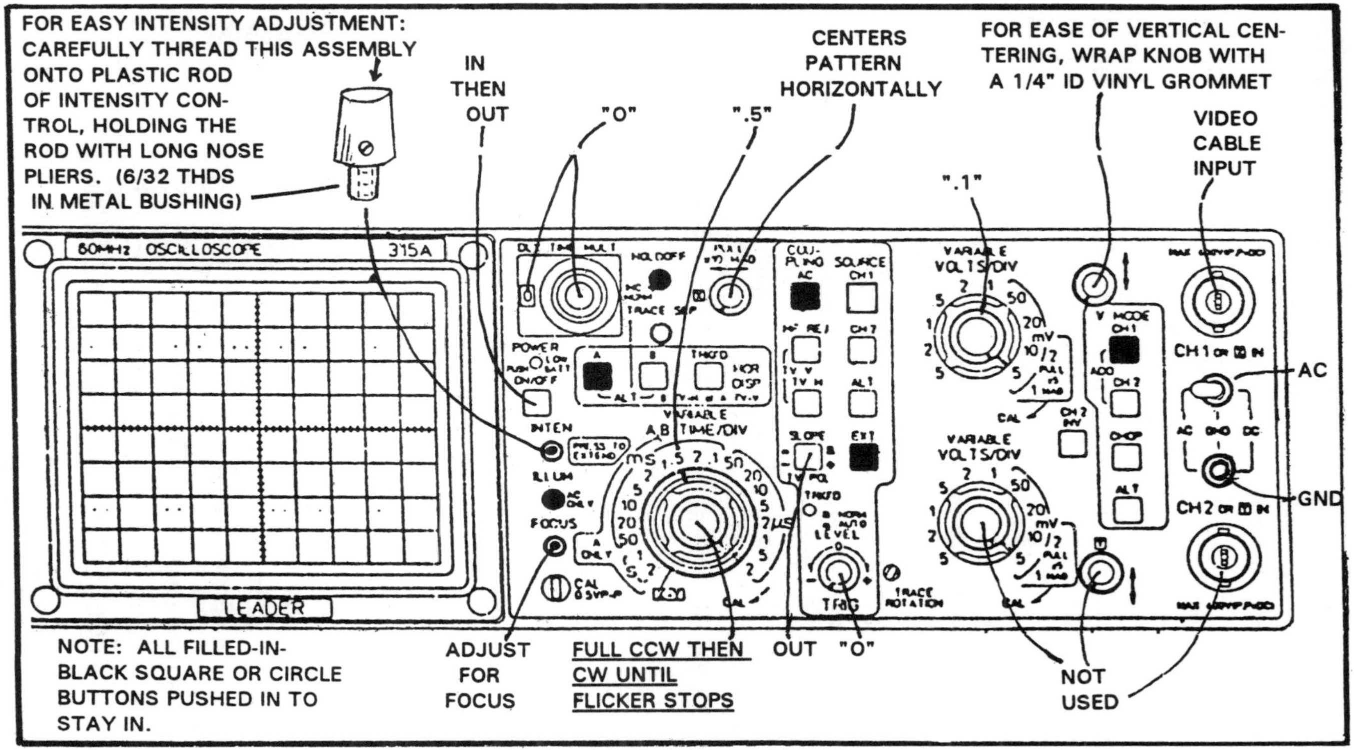

Fig. 16-3. 321-A Leader Scope: Control Setting for Interference Locating from Vehicle.

Notes, The Rotatable Antenna Mast. If you are using the Radar Engineers rotatable mast assembly and it does not have locking screws for the PVC busing nut (4) and conduit adapter 13 of Fig. 16-4, it will be worth your while to install such screws in these "O-ring adjusters."

Drill and tap these parts for 10-32 screws at the locations shown in the figure. For the top part (4), install a slotted-head brass screw of 5/16" length. For the bottom part (13A) take a 1" brass screw, cut off the head, and on that end install a radio-type control knob.

After you have installed the mast you will adjust the top part (4) and lock it down with the slotted screw. You would lock it down after you have applied silicon spray to the mast and adjusted the bushing nut for a sliding fit; after that you will adjust it infrequently.

For the inside-the-vehicle adapter (13A), you will want to adjust the tightness of the bottom O-ring occasionally, and this is the reason for a screw with a control knob. Without such a locking screw the adjuster may rotate as you rotate the mast, thereby tightening or loosening the adjustment. The brass locking screws will not damage the steel threads of the bushing (5).

As seen in Fig. 16-4, a bushing nut (4) is used for the top O-ring adjuster, and for part (3) you should use a 1-5/16" ID x 1-l/2" OD O-ring for smooth operation.

Mast Construction. If you should wish to make your own rotatable mast assembly for your vehicle, the details are shown in Fig. 16-4, and the parts list with some construction remarks follow.

Construction and Parts List, Rotatable Mast:

1. 1" PVC plastic water pipe, 200 PSI, 1-1/4" ID x 1-5/16" OD.
2. Aluminum tubing, 1-1/8" OD, .05" wall thickness. Note this tubing is used as a stiffener for the PVC pipe. It extends for the full length of mast except for the top 1/4".
3. O-ring, neoprene, 1-1/4" ID, 1-1/2" OD.
4. Conduit-type bushing nut, 1-1/4" (plastic). NOTE: Nut is reamed out slightly to provide a loose fit for the PVC mast.
5. Galvanized steel water pipe nipple, 1-1/4" x 3". If 3" nipple is not available cut from a length of 1-1/4" "conduit runner."

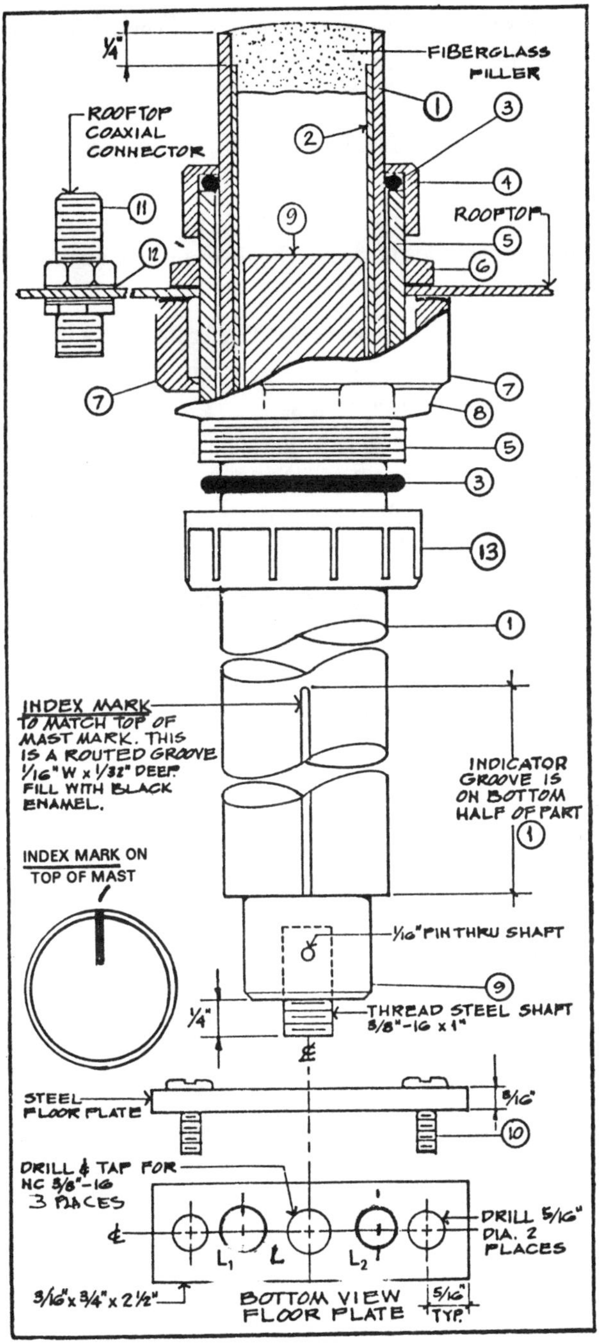

Fig. 16-4. Rotatable Mast for Locating-Vehicle. From BPA [7].

NOTE: Ends of nipple are ground square and smooth so as not to cut O-rings; inside of nipple is smoothed to prevent scoring of mast.

6. Rigid conduit sealing locknut, 1-1/4" with integral gasket, Midwest No. SL-4.
7. Spacer. Bushing nut from 1-1/2" EMT conduit compression fitting. Spacer provides increased bearing surface against rooftop for stability.
8. Rigid conduit locknut, 1-1/4". NOTE: Tighten this nut last when assembling.
9. Hardwood dowel, 11/16" diameter. This is a fixed dowel to provide stability around which the mast rotates. Length of the dowel is determined by the vehicle floor-to-roof dimension. When screwed into the floor plate, the dowel should extend up to the approximate level shown in Fig. 16-4.
10. 3/4" threaded or self-tapping screw to secure floor plate to vehicle floor.

NOTES: (1) Part 13 in the figure 16-4 should be replaced with part 13A, a 1-1/4" PVC conduit adapter, for ease of adjusting from within the vehicle.

(2) The fiberglass filler shown in the top of the mast may be replaced with a hardwood plug, cemented and pinned.

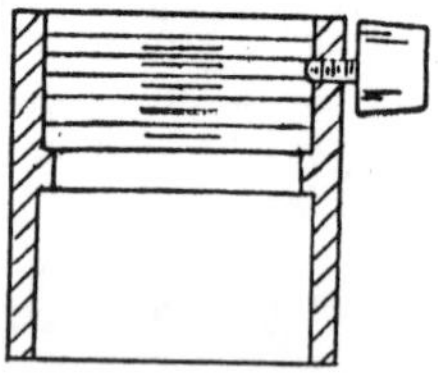

Part 13A, Section View.

(3) The floor plate should be drilled and tapped at L, L_1 and L_2 to provide for optional adjustment of mast to a vertical position for easy sliding of the mast.

Vehicular Locating Antenna. An alternative antenna for VHF "direction-finding" is described.

This is a cut-down version of the Channel Master 3619 Coloray antenna. The resulting antenna has an especially good front-to-back ratio at VHF frequencies in the 120-170 MHz range. It has a low physical profile and makes a compact package when folded up for transportation or storage. The 3619 with "cutting instructions" is shown in Fig. 16-5.

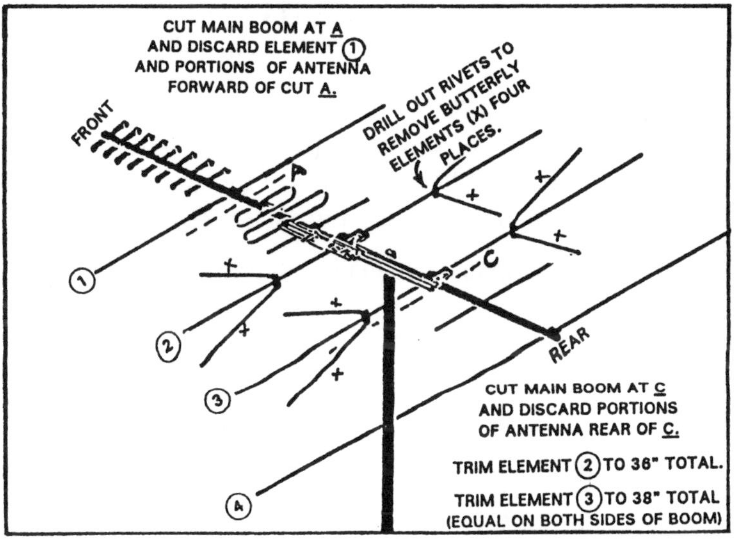

Fig. 16-5. Cutting Instructions for Making Alternate Direction-Finding Antenna for the Rapid Locating System.

After removing the "butterfly elements," you may wish to strengthen the elements 2 and 3 by slipping short lengths of thin-wall brass tubing, of snug diameter, over the rivet hole areas, and cementing or taping these lengths to the antenna elements. The author found such tubing in a hobby-craft shop.

Correction Defeated

You will find that the guy wire touching a neutral conductor is a common distribution-line source. The photo below shows an uusual and interesting example. Sparking is occurring between a guy wire and a neutral conductor. Here the guy is resting between turns of a plastic "guy-wrap." and making contact with the neutral.

This was causing TVI over a several-block area. After locating the structure with the M-330 Locator, the author, by visual inspection, observed the probable source.

As a power company person, and standing at ground level, I carefully pulled the guy away from the neutral, and the TVI in the area was cleared up.

Apparently this was causing a problem earlier, and the guy-wrap that had been installed to prevent this was defeated by the guy-wire working its way into a groove between wraps.

The power company then made a more permanent separation to resolve the problem.

Fig. 16-6. Sparking Source, Guy Wire to Neutral.

Parts Availability:

For information on obtaining parts or equivalents for the following, contact Percival Technology: percitech@att.net.

Page 7-4 The RS 12-2203 AM transistor radio for the 1-MHz Sniffer.

Page 7-7 The MM 5369 IC for the SC-3 Controller.

Page 16-9 The Channel Master 33619 antenna.

Workshops on RFI

If your company is interested in sponsoring an RF seminar, or if you wish to attend a scheduled 2-day workshop, contact: mikerfiservices@.com

A BRIEF POWER-LINE RFI GLOSSARY

<u>**AM Radio:**</u> "Broadcast radio." Non-FM radio or domestic radio, .540 to 1.600 MHz. This AM or "amplitude modulation" radio is very susceptible to noise.

<u>**Conducted Noise:**</u> RFI that travels through or along wires or via other metal pathways.

<u>**Corona:**</u> A <u>partial</u> electrical breakdown of air (or other dielectric) which radiates RFI. On EHV lines, corona can cause TVI in foul weather.

<u>**Diagonal Stripes:**</u> CB or amateur radio interference. TV set overload as result of too much booster amplification. Defective booster amplifier. Defective TV. VCR problem.

<u>**EM:**</u> Electromagnetic.

<u>**EMI:**</u> Electromagnetic interference. Also means RFI.

<u>**EHV Lines:**</u> Extra High Voltage Lines. Power lines operating at 345 kV or above.

<u>**Field Strength Meter, FSM:**</u> An RF receiver with a panel or connected meter for observing the level (or strengths) of TV or radio signals or noise. The readings indicate relative level and/or microvolts; the meter often has decibel scale. Sometimes called an SLM (signal level meter).

<u>**FM Radio:**</u> VHF frequency-modulation broadcast radio at 88-108 MHz. FM-type modulation is relatively insensitive to noise. Located in a broad band between TV channels 6 and 7, along with aircraft and other communications.

<u>**Fringe Area:**</u> Where TV signal is <u>below 225 microvolts</u> per meter (μV/m); this is outside of FCC service contour areas.

<u>**Grade A Service Contour, FCC:**</u> An enclosed geographical area where TV signal level received is <u>above 5,000 μV/m</u> for channels 2 through 6.

<u>**Herringbone stripes:**</u> See above, Diagonal Stripes.

<u>**Horizontal Bars on Screen.**</u> Co-channel interference, or radio amateur interference.

Horizontal Noise Bands on TV: A usual symptom of electrical interference on a TV–NOT necessarily power line. Can be a thermostat or a controller making noise in a building.

Induced Noise: RFI that travels in one wire, jumps to another via magnetic fields surrounding them, and continues to travel in the second wire.

Leakage Path: A path for power line current to leak to ground, to neutral, or to another phase conductor. This is usually a very high resistance path via insulators, other componentys and wood poles and cross arms.

Microspark: Term often used to mean a spark that is small compared to a power arc.

Minimum Useable TV Signal: About 35 microvolts received on a simple dipole antenna at a proposed antenna site. This requires a low-noise preamplifier and a high-gain antenna for good reception.

Noise: Interference on radio or TV or other sources. May be caused by power line or non-power line source.

Noise Bands: Dots and/or dashes sparkling on the TV screen. If drifting up across screen in two bands, it is 60-Hz power line or other electrical interference.

Noise Reradiation: *Reradiation* defines an important concept in locating power-line sparking sources. Radiation occurs at the source structure, and reradiation resulting from this specific source occurs at adjacent and other structures, and on the line.

Noise Source: Location of spark gap or other condition causing interference–usually a defect of some sort–where sparking is taking place.

Non-Power Line Source: Source located elsewhere than the power line, but often connected to it via service lines–such as a defective thermostat in a home. Sometimes called "appliance noise," "electrical Noise," or "utilization device" noise.

Power-Line Noise Source: Source located on power lines or systems.

Propagation: RFI (noise) travel by means of any of the above (radiation, conduction, induction, or reradiation).

RFI: Same as EMI. Radio frequency interference. Includes both radio and TV interference.

RI: Radio interference. Interference to AM radio, amateur, and CB radio, and to other communications services.

Signal: Desired information such as radio or TV transmissions; as opposed to "noise" which is undesired information.

Signal-to-Noise radio: The measurement of a TV or radio or communications signal level compared to a measurement of noise level as seen on the receiver-panel meter. This ratio should be about 100-to-1 or good TV reception and about 15-to-1 for good AM radio reception.

Snow: A grainy picture effect caused by weak signal. Not much detail in the picture. **Bands of Snow:** Term sometimes used for electrical interference, but it is usually not a good description.

Sparking: (Same as Microsparking). A complete electrical breakdown of air (or other gas), which generates EMI from below 1MHz to above 1000 MHz and is the cause of most RI and TVI originating on power lines. Usually is a small arc in a small gap (0.1" or less). Does not maintain an ionized conductive channel between half-cycles of voltage wave as do power arcs.

TVI: Television interference.

Vertical Stripes: A symptom of ghosting. Sometimes symptom of CB or amateur radio interference.

INDEX OF SUBJECTS